Plants, Power, and Profit

Plants, Power, and Profit

Social, Economic, and
Ethical Consequences
of the New Biotechnologies

Lawrence Busch, William B. Lacy,
Jeffrey Burkhardt, and Laura R. Lacy

Basil Blackwell

First published 1991
Basil Blackwell, Inc.
3 Cambridge Center
Cambridge, Massachusetts 02142, USA

Basil Blackwell Ltd
108 Cowley Road, Oxford, OX4 1JF, UK

Library of Congress Cataloging in Publication Data

Busch, Lawrence.
    Plants, power, and profit : social, economic, and ethical consequences of the
  new biotechnologies / Lawrence Busch, William B. Lacy, Jeffrey Burkhardt, and
  Laura R. Lacy
        p.   cm.
    ISBN 1-55786-088-2
    1. Biotechnology industries.   2. Biotechnology industries—Social aspects.
  3. Biotechnology industries—Moral and ethical aspects. I. Lacy, William B.,
  1942–   . II. Burkhardt, Jeffrey.   III. Lacy, Laura R., 1943–   . IV. Title.
  HD9999. B442B87   1990
  338. 4'76606—dc20

British Library Cataloguing in Publication Data

A CIP catalogue record for this book is available from the British library.

Typeset in 10 on 12 pt Sabon
by Graphicraft Typesetters Ltd., Hong Kong
Printed in Great Britain by T. J. Press Ltd, Padstow, Cornwall

Contents

Acronyms

ARS	Agricultural Research Service
ASTA	American Seed Trade Association
bGH	Bovine growth hormone
B.t.	*Bacillus thuringiensis*
CAST	Council for Agricultural Science and Technology
CEC	Commission of the European Communities
CHA	Chemical hybridizing agents
CIMMYT	Centro Internacional de Mejoramiento de Maiz y Trigo (International Maize and Wheat Improvement Center)
CMS	Cytoplasmic male sterility
DNA	Deoxyribonucleic acid
EEC	European Economic Community
EPA	Environmental Protection Agency
FDA	Food and Drug Administration
FIFRA	Federal Insecticide, Fungicide, and Rodenticide Act
FNWA	Federal Noxious Weed Act
FPPA	Federal Plant Pest Act
FTE	Full-time equivalent
IAC	Industries Assistance Commission
IARC	International Agricultural Research Center
ICRISAT	International Crops Research Institute for the Semi-Arid Tropics
IPRI	International Plant Research Institute
MITI	Ministry of International Trade and Industry
mRNA	Messenger Ribonucleic Acid
NAS	National Academy of Science
NASULGC	National Association of State Universities and Land-Grant Colleges
NIH	National Institutes of Health

NRC	National Research Council
NSF	National Science Foundation
OECD	Organisation for Economic Cooperation and Development
OSTP	Office of Science and Technology Policy
OTA	Office of Technology Assessment
PGI	Plant Genetics Incorporated
PPA	Plant Patent Act of 1930
PVPA	Plant Variety Protection Act
rDNA	Recombinant deoxyribonucleic acid
RNA	Ribonucleic acid
SAES	State agricultural experiment station
SSD	Single-seed descent
Ti	Tumor-inducing
TMV	Tobacco mosaic virus
TSCA	Toxic Substances Control Act
USC	United States Code
USDA	United States Department of Agriculture

Acknowledgments

The material in this volume is based upon work supported by the Ethics and Values Studies Program at the National Science Foundation and the National Endowment for the Humanities under Grant No. RII-8217306; the Agricultural Cooperative Service of the United States Department of Agriculture under cooperative agreement 58–3J31–5–0027; and the Kentucky Agricultural Experiment Station. We would like to thank the agencies that supported various portions of the research on which this volume is based. However, any opinions, findings, conclusions, or recommendations expressed in this publication are those of the authors and do not necessarily reflect the views of these agencies. As is generally true of volumes such as this, the work presented is the product of many persons in addition to the authors. Over 200 scientists and administrators in universities, private corporations, and agricultural cooperatives gave generously of their time, knowledge and insights. Michael Hansen, a plant ecologist and postdoctoral fellow in the initial stages of this project, provided important insights, particularly for the work on tomatoes, identified relevant literature, and conducted several interviews with scientists and research administrators. He also contributed to chapter 3. Glenn Collins, James Christenson, Dana Dalrymple, William H. Friedland, Joan Gussow, David Hildebrande, Richard Lewontin, Richard Merritt, Calvin Qualset, Charles Rick, Henry Shands, John Straus, and Louis Swanson read and commented on various parts of the manuscript at length. Manuel Moreira helped to locate difficult-to-obtain information on the Portuguese tomato industry. William Mesner provided several photographs from his collection and performed the photographic work on the journal advertisements. Jill Jensen has been particularly helpful in correcting and modifying the style, grammar, spelling and idiosyncrasies of the four authors, and in typing the manuscript with speed and accuracy. Nancy Strang edited the manuscript with care. To each of these persons we owe a debt of gratitude. Of course, the usual caveats apply: we alone are responsible for the errors of commission or omission that this work may contain.

Parts of this work have appeared in different forms in the authors' previously published work in the journals *Agriculture and Human Values*, *BioScience*, and *The Rural Sociologist*, and the books *Biotechnology and the New Agricultural Revolution* (J. Molnar and H. Kinnucan, eds), and *Biotechnology and the Food Supply* by the Food and Nutrition Board of the National Academy of Sciences.

1

Emerging Trends and Issues in Biotechnology

The world economy is currently undergoing major structural changes. A central factor in these changes has been the development and diffusion of fundamentally new kinds of technologies, in particular, computers and the "new biotechnologies". Social and economic changes that result from these profoundly enhanced capacities in science and technology are visible in every sphere of human life from health, transportation, and communication, to agriculture and the food supply. However, each change is associated not only with new benefits but also with new risks, latent complications and long-term consequences which are often poorly understood.

This volume focuses on the new biotechnologies, their effects on the structure of agriculture, and their impacts on agricultural science in both the public and private sectors. A broad definition of biotechnology may include any technique that uses living organisms or parts of organisms to make or modify products, to improve plants or animals, or to develop microorganisms for specific uses. However, in this book, we will focus on a limited set of new biotechnologies that primarily use novel biological methods, such as recombinant deoxyribonucleic acid (rDNA) techniques, cell fusion techniques (especially for the production of monoclonal antibodies), and new bioprocesses for commercial production (Office of Technology Assessment [OTA], 1988).

Our particular focus is on the plant sciences and the impact of these new biotechnologies on plant breeding, agricultural science, and the structure of agriculture. Commercial applications of biotechnology in plant improvement are still in their infancy, limiting what we can say about actual impacts. At the same time, however, the very lack of applications allows us to examine the alternative paths still open to development. As Winner (1986:29) recently observed, "By far the greatest latitude of choice exists the very first time a particular instrument, system, or technique is introduced." It is at this moment that a study, such as the one you are about to read, can be most helpful, not only to scientists but to the public at large.

Perceptions and Predictions

Biotechnology and genetic engineering have already begun to permeate our thinking, if not our everyday lives. They are also not without controversy, generating major debates and disagreements. According to a nationwide survey of 1,273 US adults conducted by Louis Harris and Associates, 66 percent of the public feel that they understand the meaning of genetic engineering, and more than 33 percent say they have heard or read a fair amount about genetic engineering. Furthermore, a large segment of the public (52 percent) believe genetically engineered products are at least somewhat likely to represent a serious danger to people or the environment. Moreover, 42 percent believe human cell manipulation is morally wrong, while 24 percent feel genetic manipulation to create hybrid plants and animals is morally wrong. In contrast, 66 percent of the public think genetic engineering will make life better for all people and 82 percent feel research in genetic engineering should be continued (OTA, 1987).

This tempered optimism is also reflected in the popular press. Not long ago, Wysocki (1987:1) reported in the *Wall Street Journal* that, while biotechnology is still in its infancy, "over the next few years a multibillion dollar market seems sure to develop in genetically engineered products ranging from cancer treatments to new kinds of pigs and plants". Similarly, a feature article in an airline magazine, entitled "Biotechnology: the next frontier", recently asked,

> How unknown are the mysteries of outer space? They are worn-out riddles compared to all that awaits discovery in the multiple dimensions of the New Biology.... The current focus of the New Biology is narrowing on the astonishingly fertile field of biotechnology, a scientific revolution so consequential it is changing not only how we view our world, but also how we live in the world. (Ward, 1989:85)

The *San Francisco Examiner* lead story, "Biotechnology: Man-made miracles?" recently noted that the possibilities offered by genetic engineering engender hope and optimism as well as ambivalence (Lehrman, 1989). Lehrman observed that people often refer to biotechnology as a historical development akin to the Industrial Revolution and indicates that "we are entering the Age of Biology, a time when biological sciences will become tools to solve global problems such as pollution, hunger and disease" (1989:1). She concluded, however, that while it is clear that genetic engineering can be highly beneficial to humans, the social and environmental side effects are uncertain.

The scientific community is equally positive about biotechnology's potential. Dr John Burris, Executive Director of the Commission on Life Sciences of the National Academy of Sciences, noted, "I know it's an overused term, but what we are seeing is, indeed, a revolution" (Ward, 1989:85). The Commission on Life Sciences released a report in 1989 on the new opportunities emerging in biotechnology and biological research and emphasized the revolutionary fervor in the nation's laboratories and across the country. Similarly, Dr Philip Abelson (1988), in a *Science* editorial, observed that virtually every developed country and many developing countries have targeted leadership in biotechnology as a national goal.

These predictions and perceptions are matched by those of others in government and industry. The view of the OTA is that the US is "at the brink of a new scientific revolution that could change the lives and futures of its citizens as dramatically as did the Industrial Revolution two centuries ago and the computer revolution today. The ability to manipulate genetic material to achieve specified outcomes in living organisms (and in some cases their offspring) promises major changes in many aspects of modern life" (OTA, 1987:9). Not surprisingly, Dr Howard Schneiderman, Senior Vice President for Life Sciences Research at Monsanto, indicates that this new biotechnology is "absolutely a global market" and one that some experts believe will reach $100 billion in sales by the twenty-first century (Schrage and Henderson, 1984).

Statements about agricultural biotechnology are equally optimistic. Current estimates for agricultural markets for the year 2000 range from $2–4 billion to $67 billion (Fowler et al., 1988; Organisation for Economic Cooperation and Development [OECD], 1988). Recently, the president of a large US public university noted, "Our society has moved into the era of high technology.... Before the turn of the century, we will see more technological changes than we have experienced over the entire history of our nation. It promises to be one of the most exciting and challenging times in the history of mankind." He further observed, "Biotechnology, genetic manipulation and engineering research will have tremendous impact on the crops and animals we grow for food, affecting agriculture in ways never before dreamed possible" (Lacy and Busch, 1989).

Corporate and government leaders are also positive about the potentials for agricultural biotechnology. A statement issued by Monsanto (1989) indicated that biotechnology would revolutionize farming in the future with products based on nature's own methods, making farming more efficient, more reliable, more environmentally friendly, and more profitable for the farmer. Moreover, plants would be given the built-in ability to fend off insects and disease and resist stress, animals would be born vaccinated, pigs would produce leaner meat and grow faster, cows would produce milk more economically, and food crops would be more nutritious and easier to process. The Monsanto statement concluded that biotechnology offers the American farmer an opportunity to retain the competitive edge by producing higher-quality crops and livestock at lower costs.

John Hardinger, Director of Biotechnology at the Agricultural Products Department of E. I. du Pont de Nemours views biotechnology not only as a force to restructure farming but also to catalyze a major change in the structure of worldwide agribusiness. He notes that the application of molecular biology permits the various segments of the world's largest industrial sector to form logical linkages with other economic sectors that were never before practical. The $1.3 trillion sector of agribusinesses concerned with food (not including feed and fiber) consists of four basic elements: input suppliers, growers, processors, and customers. The system has experienced mechanical and chemical eras which have contributed to increased productivity and efficiency. According to Hardinger, during the new biological and biotechnological era we will increase efficiency and productivity as well as provide the ability to change the quality of food and feed. Furthermore, we will see consolidation and new forms of vertical integration in the food industry. Hardinger concludes that "consolidation within the agribusiness sectors would be occurring even if no one had ever heard of biotechnology but biotechnology is providing one more reason for it to occur" (quoted in Klausner, 1989:219).

Similarly, OECD, an alliance of 24 countries from Western Europe, North America, Japan, Australia, and New Zealand also views the impacts of agricultural biotechnology in terms of major product and structural changes. OECD notes a probable increase in the convergence of agricultural and industrial practice, and predicts a reorientation in relationships between agricultural suppliers, farmers, and the food processing industry with a new generation of science-based agricultural companies designed to exploit these technologies through products destined for a variety of markets. The agricultural sector may become increasingly reliant on high technology, and biotechnology may affect agricultural surplus, trade, employment, alternative land use, and ecological systems (OECD, 1988).

Biological Developments

Despite these optimistic predictions, most of the products, processes, and impacts of biotechnology, particularly in agriculture, remain promises for the future. Even in the pharmaceutical industry, where biotechnology has been an integral part of research for over a decade, only seven human therapeutics using biotechnology were approved for marketing in the US by early 1988 (OTA, 1988). Of 25,000 active applications for investigational new drugs, only 400 (less than 2 percent of the total) biotechnology-based human therapeutics were in some stage of clinical trials. However, human health care, primarily therapeutics and diagnostics, continues to be the focus of most biotechnology research and development investments, with agriculture, as a field of application for industrial biotechnology, remaining a distant third. Indeed a recent OECD report entitled "Economic and wider impacts of biotechnology" concludes that not until the turn of the century will biotechnology begin to play an economic and social role comparable to that of information technology, and only in the second decade of the century will it have major macroeconomic impacts (Dixon, 1989).

Nevertheless, over the last 15 years there have been a series of important discoveries and dramatic developments that have emerged from basic research in molecular biology. After the first successful directed insertion of foreign DNA into a host microorganism in 1973, researchers recognized the opportunities for advances in basic molecular biology, and the commercial potential in such areas as pharmaceuticals, animal and plant agriculture, specialty chemicals and food additives, commodity specific chemicals, energy production, and bioelectronics. This development was quickly followed in 1974 by the expression in bacteria of a gene cloned from a different species. Other major biological developments in the last decade have included the first rDNA animal vaccine approved for use in Europe (1981), the approval of the first rDNA pharmaceutical product (human insulin) for use in the US and the United Kingdom (1982), the first expression of a plant gene in a plant of a different species (1983), and the 1988 transfer of genes from other species into mice creating the first transgenic animal (Wheeler, 1988). Finally, in mid-1989, researchers at the National Institutes of Health (NIH) introduced the first foreign gene into a human patient in an experiment designed to develop a new way of treating malignant melanoma (Associated Press, 1989).

Animals

In agriculture, the impacts of molecular biology and biotechnology may entail improvements and modifications in the traditional means of production of animals, plants and food products. Since World War II, the animal sciences have benefited from a molecular biological emphasis with the results of medical research often directly transferrable to meat and dairy animals. Biotechnology is likely to enhance livestock production through better diagnostic products, vaccines, growth promotants, and manipulation of the animal genome to achieve expression of desired traits. Among the recent developments are a genetically engineered vaccine against scours (a yearly killer of millions of newborn calves and piglets), a subunit vaccine engineered to prevent foot-and-mouth disease, and the pseudorabies vaccine (Baumgardt, 1988). Other animal science research focuses on lowering the fat content of farm animals by a variety of methods, including genetic engineering of the livestock. Work has also been conducted on growth hormones produced by rDNA technology to increase muscle growth more efficiently and reduce fat production without compromising nutritional value (Sun, 1988). Although linked to improvements in daily weight gain and feed efficiency as well as marked reduction in subcutaneous fat, long-term elevation of bovine growth hormone (bGH) in two lines of transgenic pigs was found to be associated with several negative effects including gastric ulcers, arthritis, cardiomegaly, dermatitis and renal disease (Pursel et al., 1989). Related studies have evaluated the effects of bGH on milk yields in dairy cattle. In certain climates, tests show increases in both milk production and feed efficiency with proper management. However, such developments are not always enthusiastically greeted. At a time of great overproduction of milk, Wisconsin dairy owners fear larger surpluses, plunging prices, and further loss of dairy farms. Additionally, many consumers are raising questions about the quality and safety of milk produced with the use of bGH. Both groups have called for the Food and Drug Administration (FDA) to refuse a commercial license for bGH use (Sun, 1989). These promising technical developments illustrate that even the best intentions and apparently successful genetic engineering can produce a range of negative biological and social effects (Lacy and Busch, 1988).

Plants

Despite the early advances in animal science, many analysts are predicting that the new biotechnologies may have their greatest impact on plants. Monsanto scientists note that use of genetically engineered plants as an analytical tool for exploring unique aspects of gene regulation and development with the potential for producing new commercial crop varieties, has generated much excitement among scientists and has propelled research into many new areas (Gasser and Fraley, 1989). In the plant sciences, current applications of biotechnology emphasize modification of specific plant characteristics (e.g., resistance to weeds, pests, herbicides, and pesticides; tolerance to stress; and nutritional content), and traits of microorganisms that could be important to plant agriculture (e.g., those that foster pest resistance, nitrogen fixation, frost resistance, and disease suppression). Initial research has been focused more narrowly on genetically engineered traits that relate directly to such traditional approaches as control of insects, weeds, and plant diseases. The short-to-medium term commercial strategies for farms concentrate on market expansion or extension for registered crop chemicals through promotion of herbicide tolerance and development of encapsulated embryos to replace traditional seed. Longer-term corporate and university strategies assume some shift away from synthetic crop protection toward biological pest controls and fertilizers (e.g., bioinsecticides, bioherbicides, biofungicides, and nitrogen fixation) (Fowler et al., 1988) and the development of genetically engineered hybrids. For example, an important achievement that was recently announced was the first successful demonstration that maize plants possessing a novel gene could be stable, fertile, and capable of transmitting the gene to the next generation (Genetic Engineering News, 1990b). Furthermore, a gene that prevents pollen development has recently been expressed in crop plants. This will allow seed companies to create crop hybrids that are not technically and/or economically feasible through conventional methods (Genetic Engineering News, 1990a).

A number of important preliminary developments have occurred in the methods for introducing genes into plants, including technologies for regulating gene expression and isolating genes (e.g., introduction of regulatory and antisense sequences, transposon-mediated tagging, and high resolution restriction fragment length polymorphism gene mapping) and the application of genetic engineering to crop improvement. Through several techniques, such as protoplast fusion, clonal propagation, and direct transformation (primarily through the use of the *Agrobacterium* system), important genes have been transferred to commercially valuable crops. In addition, a variety of free DNA delivery methods, including microinjection, electroporation, and particle gun technology, are being developed for the transformation of monocots such as corn, wheat, and rice (Gasser and Fraley, 1989). Transgenic plants have been produced in over three dozen species including herbaceous dicots such as tobacco, soybeans, cotton, carrots, tomatoes, potatoes, alfalfa, oilseed rape, and sunflower; woody dicots such as walnut and apple; and monocots such as rice, corn, and rye. Researchers at Monsanto predict that genetically engineered soybean, cotton, rice, corn, oilseed rape, sugarbeet, tomato, and alfalfa crops will enter the market place between 1993 and 2000 (Gasser and Fraley, 1989).

Weed control   Most of the research in the area of weed control centers on herbicide tolerance. At least 27 enterprises have research programs directed toward the development of herbicide tolerant or resistant crop varieties, with a market value expected to exceed $6 billion by the turn of the century (Fowler et al., 1988). Industry researchers report that work has generally concentrated on those herbicides with properties such as high unit activity, low toxicity, low soil mobility, rapid biodegradation, and broad spectrum activity against various weeds (Gasser and Fraley, 1989). The two general approaches utilized in engineering herbicide tolerance are altering the level and sensitivity of the target enzyme for the herbicide, and incorporating a gene that will detoxify the herbicide. The first approach was employed in several food crops to engineer tolerance to glyphosate, a major herbicide marketed by Monsanto as Roundup®. As a consequence, projections for increased annual sales of Roundup®, the world's largest-selling herbicide, have been as high as $150 million (Fowler et al., 1988). Similarly, resistance to the active ingredients in Glean® and Oust® herbicides have also been produced. In an interesting example of corporate collaboration, American Cyanamid recently developed a new family of imidizolinone herbicides and contracted Molecular Genetics to identify a gene conferring tolerance to the chemical. The chemical company then negotiated with Pioneer Hi Bred, the world's largest maize breeding company, to conduct research to insert this gene into its hybrids. Other current crop targets for engineered herbicide tolerance include soybean, cotton, oilseed rape, and sugarbeet.

Insect resistance   Production of insect resistant plants is another area of new genetic engineering applications with important implications for crop improvement. However, the number of research programs is only half that for herbicide tolerance, and the scope of work is surprisingly narrow. Some analysts estimate that 95 percent of the commercial research in bioinsecticides is concentrated on the introduction of *Bacillus thuringienis* (*B.t.*) genes into crops. Strains of this bacterium produce insect control proteins which are toxic to selected insect pests including some lepidopteran pests (moths and butterflies), and certain beetle and fly larvae, while not affecting beneficial insects (e.g., honeybees and ladybugs), other animals, or humans. Transgenic tomato, tobacco, and cotton plants containing a *B.t.* gene exhibited resistance to caterpillar pests in laboratory tests (Gasser and Fraley, 1989). Annual expenditures for insecticides against lepidoptera are estimated at $1.5 billion across all crops, and commercial applications of these bioinsecticides may appear in the near future. However, recent evidence from field trials on transgenic plants indicates that *B.t.* resistance may develop in natural insect populations (Fowler et al., 1988). Consequently, transgenic plants producing mixtures of *B.t.* proteins or other types of insecticidal molecules may be required to extend biotechnology approaches for controlling these and other insects.

Disease resistance Development of disease resistant plants is another important area of crop research. Significant resistance to tobacco mosaic virus (TMV) has been achieved by expressing the protein coat of TMV in transgenic plants. This approach has produced similar results in transgenic tomato, tobacco, and potato plants against a broad spectrum of plant viruses. For example, Monsanto researchers recently announced that they had genetically engineered potatoes to resist potato viruses X and Y (PVX and PVY) for which there are no chemical controls (Lawson et al., 1990). As a consequence, significant yield protection could be provided in important crops such as vegetables, corn, wheat, rice, and soybeans. Moreover, limited success in engineering resistance to fungal diseases has been reported, though work is in the early research stages. This work is particularly important since fungi are the largest and most economically important group of plant pathogens, causing billions of dollars of crop losses and postharvest food spoilage each year (Timberlake and Marshall, 1989). Current world spending on fungicides is estimated to be $2.3 billion.

Biofertilizers Finally, in the area of biofertilizers, progress has also been made in increasing the nitrogen-fixing capacity of some leguminous plants. Biotechnica International has developed a modified bacterium (*Rhizobium melitoti*) that may increase the yield of alfalfa by 15 percent. If field tests proceed on schedule, improved *R. melitoti* will be available to farmers as a soil inoculant in the early 1990s. According to a company representative, this genetically engineered bacterium could be worth up to $1 billion to the nation's alfalfa growers (Sterling, 1988). Another effort that resulted in an increase in biomass production of more than 10 percent in the respective hosts under greenhouse conditions involved modifying the expression of two specific genes in nitrogen-fixing symbionts (Lindow et al., 1989). In other developments, US scientists at three universities have successfully transferred nitrogen-fixing genes to non-nitrogen-fixing crops. The symbiotic associations among the crops and the various genes from the bacteria (e.g., *Rhizobium, Bradyrhizobium,* and *Frankia*) are highly complex but offer many important future possibilities. Scientists believe that ultimately it may be possible to manipulate both crop and rhizobial genes to achieve maximum efficiency of nodule formation and function, and to tailor strains for unusual soil environments (Lindow et al., 1989). If nitrogen-fixing genes transferred to other crops prove to be commercially viable, the reduction in the use of nitrogen fertilizers could be substantial. On the other hand, nitrogen fixation research could lead to the development of plant varieties capable of utilizing synthetic fertilizers more effectively or absorbing greater quantities of them, thereby increasing production costs and environmental hazards.

Food Processing

The third major area of the food and agriculture system where biotechnological developments are already having an impact is food processing. Here the techniques of growth and fermentation of yeasts and bacteria – well known in certain segments of the food processing industry, especially in cheese and bread making and alcohol production – will be equally important in the future. Genetic engineering is already being applied to bacteria, yeast, and fungi to produce starter cultures with specific metabolic capabilities in food fermentation. Other areas of application of genetic engineering include plant tissue culture technology for the production of plant-derived food ingredients; protein engineering to tailor enzymes with specific catalytic capabilities (including thermostability, substrate specificity, and pH stability) so they are more amenable to processing systems; and genetically modified microorganisms to enhance production of high-value food ingredients to facilitate waste management in the food industry and to improve food safety (Newell and Gordon, 1986). These processes, combined with the new cell culture techniques, are being used to transform the production of certain agricultural commodities into industrial processes. In principle, any commodity that is consumed in an undifferentiated or highly processed form could be produced in this manner, and product substitutions could easily be introduced. Similarly, although with greater difficulty, tissue culture techniques could be used to produce edible plant parts *in vitro*. In short, agricultural production in the field could be supplanted by cell and tissue culture factories.

Several examples have emerged. Companies based in the US are now attempting to produce a natural vanilla product in the laboratory through phytoproduction. Escagen, a small California biotechnology company, marketed a laboratory-produced natural vanilla in 1989 (Fowler et al., 1988). The genetic modification of oil seed plants to convert cheap oils (e.g., palm or soybean oil) into high-quality cocoa butter is well advanced. A food processing magazine notes that to avoid perceived problems of importing oils, the industry will produce similar oils from domestic sources and even create oils not found in nature. Another example is the application of biotechnology to produce substitutes for sugar as an industrial sweetener. Several major corporations in the US and Europe, such as Unilever and Ingene, are attempting to use rDNA technology to produce the thaumatin protein, one of the sweetest known substances which occurs naturally in a West African plant. The development of a thaumatin product through genetic engineering may continue a transition to alternative sweeteners eliminating the market for beet and cane sugar and capturing the valuable sweetener market, currently worth $8 billion in the US (Fowler et al., 1988).

Environmental Concerns

Accompanying these biological developments are increasing concerns about the environmental and ecological impacts of biotechnology. Many groups of scientists and environmentalists, and many local citizens' groups, have raised safety issues regarding the unexpected (but possible) consequences of introducing new organisms, such as the production of a toxic secondary metabolite or protein toxin, or undesired self-perpetuation and spread of the organism (Brill, 1985). In agriculture, some have suggested that living natural inputs may be even more dangerous to society than the artificial products they replace (Fowler et al., 1988). Robert Goodman, Vice President of Calgene, one of the leading agricultural biotechnology companies, notes that some of nature's allopathic chemicals include arsenic and cyanide. "Just because it's natural doesn't mean it's safe" (quoted in Fowler et al., 1988:85). Critics of the research, however, fear unintended and perhaps irreparable damage to natural ecosystems by transgenic organisms that nature could never create (Witt, 1990). In addition, there is concern that genetically engineered crops bearing resistance to major herbicides may become weeds. Furthermore, some fear that these herbicide-resistant crops may cross with weedy relatives and spread resistance into sectors of the weed flora (Klassen, 1989). These critics have successfully sought court injunctions to suspend work and to prevent research that entails the release of new microbes in a natural setting. Finally, extensive governmental mechanisms for review and approval of biotechnological research have been developed in an effort to address some of these important concerns.

Government Biotechnology Investment

Federal

The rapid biological developments described above have been made possible in part by major financial investments in basic and applied research by federal and state governments in the US. Twelve federal agencies and one cross-agency program spent $2.7 billion in fiscal year 1987 to support biotechnology research and development. This support has increased every year since 1984. The largest share of support was provided by NIH, nearly $2.3 billion (84 percent), while the Department of Defense contributed $119 million, the National Science Foundation (NSF) $94 million, the US Department of Agriculture (USDA) $84 million, and the US Department of Energy $61.4 million. Nearly 60 percent of these funds was earmarked for basic research (OTA, 1988).

However, in contrast to federal support of biomedical research through NIH, funds for basic research in agriculture and plant biology have been meager, totaling approximately 3 percent of the federal biotechnology budget. In addition, only 1.4 percent of USDA's budget is devoted to research, with less than 5 percent of that amount earmarked for biotechnology. In 1985, an effort was made to increase agricultural biotechnology research with the addition of $20 million to the USDA's Competitive Research Grants Program. However, by 1989, the funding for this biotechnology program had actually dropped 5 percent to $19 million (Stumpf, 1989). The situation appears slightly better when funds from all federal sources devoted to agricultural biotechnology are summed, giving a total of $150 million (OTA, 1988). However, Abelson (1988), Deputy Editor of *Science*, notes that, while the US is a leader in biotechnology in pharmaceuticals, it could become second-rate in agriculture. Among the reasons he cites are the low levels of federal funding necessary to build the knowledge base of plant molecular biology. The National Research Council's (NRC's) Board on Agriculture has identified the same issue and has recommended that federal funding in agricultural biotechnology be increased to $500 million annually by 1990 (Moses et al., 1988).

State

In the US, allocations in biotechnology are increasing among 33 states which have actively engaged in some form of promotion of biotechnology research and development. States have allocated funds through centers of excellence, university initiatives, and incubator facilities for new firms or for grants for basic and applied research in biotechnology. In 1987, state biotechnology investments designed to achieve academic excellence in the state's universities, and/or economic development, totaled $147 million. The states employ various funding mechanisms including public bonds, allocation of state lottery funds, mandatory industry–university collaboration, and matching funds. Much of this investment, constituting more than half the total, is concentrated in just three states: New Jersey, New York, and Pennsylvania (OTA, 1988). A substantial portion of these funds is being devoted to agriculture and is invested in new centers and institutes. Some examples are the New York State Center for Biotechnology in Agriculture at Cornell University (1983), the Center for Agricultural Molecular Biology at Cook College, Rutgers University (1987), Iowa's Office of Biotechnology at Iowa State University (1984), and the Biotechnology Institute at Pennsylvania State University (1984). Most of these programs are too young to evaluate but expectations for their success run high.

Foreign

Several foreign governments are also investing in biotechnology. While their support is substantial, it is considerably less than that of the US government. Unlike the US, where there is an emphasis on basic research, the Japanese and many European governments have focused on the commercial goals of biotechnology and have supported applied research and development efforts that emphasize the transfer of research from government laboratories to industry. Japan's Ministry of International Trade and Industry (MITI), which has guided development efforts in automobiles and computers, has designated biotechnology as a strategic industry and a national priority. Under MITI, groups of companies have been established and seed money has been provided in four areas of biotechnology: gene-splicing, large-scale cultivation of cells, bioreactors or microbes that change one chemical into another, and bioelectronics (Wysocki, 1987). Research and development support for Japanese firms has been boosted to several hundred million dollars; in 1986, the government's biotechnology budget was $196 million. It is estimated that this strong Japanese emphasis on commercial development of biotechnology will account for over 10 percent of the Japanese gross national product by the year 2000 (Ward, 1989).

Of the European nations, the United Kingdom is the most involved in biotechnology. For 1989–92, the United Kingdom has allocated an additional $100 million of public funds for biotechnology-related research to strengthen its initial investments. An additional $25 million will be provided to the Agricultural and Food Research Council for plant molecular biology (Newmark, 1989). Four other European nations – Spain, the Netherlands, France, and West Germany – also have national programs to support and coordinate biotechnology efforts. In West Germany during the mid-1980s, three federally sponsored biotechnology research institutes joined with industry to invest an estimated $280 million in biotechnology research (OTA, 1984). In addition to the efforts of individual nations, the Commission of the European Communities (CEC) has budgeted approximately $150 million for biotechnology for 1987–91.

Universities

Other major contributors to the development of biotechnology are universities. Traditionally, universities have been an important component in state, regional, and national development and problem solving. Indeed, biotechnology owes much of its growth to academic science.

Industry has looked to universities as sources of cutting-edge research and collaborative work, while states have viewed their universities as the bases of their biotechnology efforts and as a means of attracting biotechnology companies. Of the 33 states reporting state-supported biotechnology programs in the US, 28 primarily relied on higher education institutions for the design and performance of biotechnology. Moreover, in 14 states, universities have been the impetus for creating biotechnology programs (OTA, 1988).

A rich variety of interdisciplinary centers, institutes, research parks, and corporations have been created at American universities to pursue biotechnology. In the area of agricultural biotechnology, there are at least 17 state university research center initiatives located at land-grant universities. These new research centers focus on the major growth areas of biotechnology and tend to be larger and better financed than their precursors. In addition, they have generally involved joint funding, performance, and application of scientific work by universities and industries. While partnerships between universities and industries have existed for several decades, the new types of university–industry relationships in biotechnology are generally more varied, wider in scope, more aggressive and experimental, and higher in public visibility than the relationships of the past (Lacy and Busch, 1989).

During the 1970s, several key factors in the university and industrial environments stimulated universities to seek greater private sector support. These included increasing research costs which exceeded available funds from traditional sources; increasing federal budget deficits and the change of administration in 1981 which prompted research administrators to fear science budget cuts; and the recognition by industry that US technological supremacy was eroding and being challenged. These changes were coupled with government patent, tax, and funding policies to strengthen the links between industry and university research (e.g., Patent and Trademark Amendments Act of 1980, Stevenson–Wydler Technology Innovation Act of 1980, Economic Recovery Tax Act of 1981, and Technology Transfer Act of 1986) (OTA, 1988; Booth, 1989).

The relationships between universities and industries depend on the goals and institutional characteristics of the partners, and consequently encompass diverse approaches. These biotechnology arrangements – between firms, universities, faculty members, and students/trainees – include large grants and contracts between companies and universities in exchange for patent rights and exclusive licenses to discoveries; programs and centers at major universities, funded by industries, that give private firms privileged access to university resources and a role in shaping research agendas; professors, particularly in the biomedical sciences, who are consultants on scientific advisory boards or are managers in biotechnology firms; faculty who receive research funds from private corporations in which they hold significant equity; faculty who set up their own biotechnology firms; public universities that establish for-profit corporations (e.g., Neogen at Michigan State University) to develop and market innovations arising from research; and universities that establish alumni panels with venture capital to evaluate new university technologies for possible commercialization (Fuchsberg, 1989; Lacy and Busch, 1989).

Approximately 46 percent of the biotechnology firms support biotechnology in universities. In 1984, these firms awarded about $120 million in grants and contracts to universities. Furthermore, approximately 16–24 percent of university research in biotechnology is industrially sponsored compared with 4–5 percent of overall campus research (Blumenthal et al., 1986b). In a survey of over 1,200 faculty members at 40 major universities, 47 percent of biotechnology faculty reported consulting with an outside company, and 8 percent reported holding equity in a firm whose products or services were directly related to their own university research (Blumenthal et al., 1987).

In agricultural biotechnology, a number of university–industry relationships have emerged. Cornell University has specific grants from Agrigenetics Corporation (tomato hybridization through cell cultures), Sandoz (tissue culture generation), and Nestle's (bitterness in squash); Purdue University has grants from UpJohn and Dow Chemical (soybean research), Eli Lilly (wheat genetics), and Quaker Oats (oats research); and Pennsylvania State University has arrangements with Hershey Foods, Mars, Inc, and Ambrosia Chocolate Company (molecular biology of the cocoa plant), and Pioneer Hi-Bred (grain yield in maize). Other universities with domestic corporate relationships in agricultural biotechnology include the University of Arkansas (Busch Agricultural Resources, Inc. and Cotton, Inc.), Washington University (Monsanto), the University of Pennsylvania (Rohm and Haas), and the University of Illinois (Standard Oil) (OTA, 1988). In addition, numerous American universities have established relationships with Japanese firms. These include Columbia University (Yakuit), Harvard University (Takeda), Vanderbilt University (Daiichi Seiyaku), the University of California at Irvine (Hitachi Chemical), and MIT (Ajinomoto) (Martin, 1990).

Perhaps the most advanced university–government–industry consortium to focus on agricultural applications of biotechnology is the Midwest Plant Biotechnology Consortium, which involves 15 midwestern universities, three federal laboratories, 37 agribusiness corporations with headquarters in the Midwest, and research institutes from eight states. The consortium's purpose is to conduct basic research in plant biotechnology and to promote the transfer of technology to foster the economic competitiveness of US agriculture and agribusiness. While many details and antitrust obstacles must be addressed, this arrangement shows promise (OTA, 1988).

It seems unlikely, however, that university–industry relationships will either significantly address the capital needs of universities or provide stable long-term funding. In 1984, for example, 60 percent of industry-supported biotechnology projects at universities were funded at less than $50,000, while only 20 percent were funded for over $100,000. Moreover, half of the biotechnology companies funding university research supported projects that lasted one year or less, while only 25 percent supported projects that lasted more than two years. In contrast, 92 percent of NIH's extramural awards were for projects lasting three years or longer (Blumenthal et al., 1986a, 1986b).

Finally, these relationships raise new issues for universities; for example, does corporate penetration into academic science distort traditional values of basic research and alter the atmosphere, working climate, and free communication of academic departments? Universities are also concerned about ensuring that research projects are genuinely originated by faculty members and not adopted as a result of outside pressure, either implicit or explicit. A related issue – the nature of the research agenda – has also been raised: if a sufficiently large and influential number of academic scientists and engineers become involved with industry, a whole range of research agendas, traditionally part of the university community, might be abandoned. Furthermore, the scientific community could become desensitized to the social impacts of biotechnology research. Some research that lacks commercial application could be neglected entirely (Krimsky, 1984; Lacy and Busch, 1989). In addition, the balance between short-term and long-term research agendas, traditionally a concern of the public sector, may be disrupted by the new collaborative research programs with the private sector. Another issue involves the potential conflict of interest between industry and university agendas. New research centers have divided commitments and loyalties created by joint appointments, while universities are concerned with maintaining control of appointments to these centers. A related issue that goes beyond the boundaries of a single university is that a research center can confer a high-tech advantage on a particular firm rather than serving a broader public.

Commercial Developments in Biotechnology

The emergence of a worldwide biotechnology industry parallels these rapid scientific developments and government and university investments. This industry began in the US in 1976 with the founding of Genentech (the first firm to exploit rDNA technology) by Herbert

Boyer, a University of California bacteriologist and venture capitalist. This link with US universities initiated a pattern for early commercial development; all of the earliest genetic engineering companies were founded by university professors (Kenney, 1986). *BioTechnology* magazine reported that half of all the founding fathers of biodrug companies were from public institutions (Fowler et al., 1988).

The formation of biotechnology companies boomed from 1979 to 1983, with more than 250 small, venture-capital biotechnology firms founded in the US. Generous tax incentives fueled the boom. Approximately 60 percent of the companies existing today were created during those years. The proliferation of these risk-taking companies helped raise over $4 billion from private investors and gave the US a competitive lead in the early stages of biotechnology's commercialization. The number of these firms grew to over 400 by the mid 1980s. Since then, consolidation within the industry and the predominance of a small number of corporations have slowed the formation of new firms. Indeed, 80 percent of the funds in venture-capital firms have been invested in just ten companies; investments in health-care applications account for 75 percent of all investment and in agriculture, 16 percent (OTA, 1988).

By the early 1980s, however, larger multinational corporations began to recognize the potential of biotechnology and to develop their own research and development capacities. These corporations began diversifying into every field or specialty that used living organisms as a means of production. The new biotechnologies appear to be further reducing the distinctions among the traditional industrial sectors and rendering corporate boundaries virtually unlimited. For example, large nonpharmaceutical corporations, such as Du Pont, Nestle, Monsanto, Procter and Gamble, and Dow Chemical have developed a new emphasis on, and have invested heavily in, pharmaceutical efforts through biotechnology research and development. In 1981, Du Pont announced a $120 million research and development program in the life sciences, with a particular focus on biotechnology. During the early 1980s, Monsanto opened a $150 million Life Sciences Research Center with state-of-the-art biotechnology facilities (Fowler et al., 1988). The development of in-house expertise continued to grow throughout the 1980s. By 1986, large corporations provided over half ($2.2 billion) of the total funds for biotechnology. By the late 1980s, 75 large, established companies averaged $11 million annually for biotechnology research and development, with nearly every firm spending some of its budget in-house. Combined with the venture-capital firms these companies employed 36,000 workers in biotechnology activities, 18,600 of whom were scientists and engineers (OTA, 1988). This emerging field now includes nearly 900 companies that provide equipment, instrumentation, supplies, and services for research and development (*Bio/ Technology*, 1988a).

Collaboration

Most large corporations were initially reluctant to invest heavily in this undeveloped area. Consequently, they began by establishing strategic partnerships with small biotechnology firms and university research programs, permitting them to reduce financial risk while ensuring early access to biological developments, products, or production technology and knowledge. For small biotechnology firms, this collaboration allowed them to overcome resource limitations which could have prevented them from developing or marketing products themselves. Over four-fifths of the established large companies investing in biotechnology spent some of their research and development budgets in research conducted by outside firms and universities (OTA, 1988).

Today, a wide variety of alliances have been formed between the approximately 400 American companies dedicated to biotechnology and the large multinational corporations with significant investments in biotechnology. These arrangements include research and development contracts, equity positions, joint ventures, licenses, patents, and marketing agreements. While collaborations are not always between large corporations and small ones, OTA estimates that three-quarters of the agreements are of this nature. Moreover, there are 1,000 to 1,500 joint venture agreements between the more than 800 firms active in biotechnology worldwide, but less than one-quarter are international collaborations. In general, large corporations acquire licenses to market products but not to manufacture products. In addition, contrary to most expectations, the number of agreements is remaining stable or increasing, although this situation may change in the future (OTA, 1988).

Currently, the collaborative ventures experiencing the most intense business activity are in the areas of human therapeutics (29 percent) and clinical diagnostics (25 percent). These arrangements include both multiple alliances and nontraditional linkages. For example, by the late 1980s Genentech, the first and then largest biotechnology company in the US, had 13 US corporate partnerships, at least seven Japanese corporate partnerships, and four European partnerships (Fowler et al., 1988). By 1986, Amgen, a pharmaceutical biotechnology firm, had established partnerships with eight prominent corporate partners: Abbott Laboratories, Arbor Acres Farm, Johnson and Johnson, Kirin Brewery, Eastman Kodak, SmithKline Beckman, Texaco, and UpJohn. Similarly, Calgene, an agricultural biotechnology company, joined forces with Agrochemie, Campbell Soup, Ciba-Geigy, Kemira Oy, Philip Morris, Procter and Gamble, and Roussel-Uclaf (OTA, 1988). For

Procter and Gamble – a food company interested in reverse integration into the supply sector of agriculture, and a major marketer of oils – the equity stake in Calgene provided access to innovative work on rapeseed oil (Klausner, 1989). Interestingly, each of these biotechnology companies has managed to negotiate separate agreements with traditional competitors.

Large corporations have also signed multiple arrangements with biotechnology firms. As early as 1984, Eastman Kodak contracted with nine venture-capital biotechnology firms to work in areas as diverse as cancer drugs and genetically engineered indigo dye for blue jeans. Other collaborations include Johnson and Johnson's 11 equity positions in biotechnology companies and American Cyanamid's more than 15 licensing arrangements with biotechnology firms. Recently Du Pont, which has traditionally based its agribusiness strength in crop protection products in the supply end of the system, agreed in principle to invest over $14 million in the company DNA Plant Technologies (DNAP) for developing and marketing fruits and vegetables with superior taste and prolonged shelf life (*Bio/Technology*, 1988b). Today, these firms and their collaborative arrangements reportedly invest between $1.5 and $2.0 billion annually in biotechnology research and development (OTA, 1988).

Despite these major investments, biotechnology firms are faced with serious problems and widespread consolidation. The venture-capital companies have been hampered by delays in product development, regulation, patent delays, financial market instability, and general public concerns. Nearly all of the biotechnology firms continue to lose money. Indeed, in 1986 the US industry lost $480 million (OTA, 1988). In 1987, an OTA survey revealed that the average biotechnology firm was six years old, operated with a research and development budget of approximately $4 million, had a total income of only $10 million and did not operate at a profit. In 1987, one financial consulting firm predicted that the number of US biotechnology companies would decrease by 33–50 percent every 10 years until fewer than ten survived; this consolidation would occur equally from company failures, mergers, and acquisitions (Fowler et al., 1988). Similarly, a recent survey of biotechnology firm executives resulted in a prediction that, within 10 years, roughly half of the nearly 500 US biotechnology companies would fail, merge, or form cost-sharing alliances (Naj, 1989).

This process has already begun. In 1988, at least 24 biotechnology companies filed for bankruptcy protection, and there have been numerous mergers and acquisitions (Naj, 1989). A prime example was the 1989 announcement that the Philadelphia-based firm SmithKline Beckman and the London-based firm Beecham Group PLC would merge and become the second largest pharmaceutical company in the world, with combined 1988 revenues of $6.7 billion (*Bio/Technology*, 1989b). Another 1989 merger of biotechnology firms occurred between Genzyme, with expertise in carbohydrate chemistry and enzymology, and Integrated Genetics, with the ability to develop and manufacture genetically engineered proteins. Perhaps the most significant merger is the 1990 purchase of Genentech by Roche Holding Limited for $2.1 billion, thus making the Swiss-based pharmaceutical company the leading force in biotechnology worldwide (Gebhardt, 1990).

Foreign Developments

The investment in biotechnology research and commercialization has also occurred, albeit to a lesser extent, in several other countries (e.g., Japan, the United Kingdom, the Federal Republic of Germany, Switzerland, and France). Japan has always been recognized as a leading actor in biotechnology and its influence appears to be increasing. In contrast to the US, Japan's strength in biotechnology did not arise from newly formed small biotechnology firms, but rather from large established corporations representing virtually all industrial sectors. In 1986, 325 Japanese companies spent approximately $860 million for research and development in biotechnology. According to Wataru Yamaya, Managing Director of Japan's Mitsubishi Chemical Industries Ltd, "Biotechnology is a magic word in Japan" (Wysocki, 1987:1). His company is devoting about 40 percent of its approximately $240 million research and development budget to biotechnology.

The pattern of corporate horizontal and vertical integration generated by biotechnology in the US is also occurring in Japan. For example, at least ten major Japanese chemical companies, 15 food processing companies, and four textile companies have moved into biotechnology-related pharmaceuticals. Japanese brewers and distillers, such as Kirin and Suntory, are building on their history with fermentation technologies to become major actors in pharmaceuticals (Fowler et al., 1988). Suntory synthesized the gamma interferon gene in 1982 and is currently using gamma interferons in tests against leukemia, solid tumors, and rheumatoid arthritis. In 1987 it was reported that Suntory had plans to open a $100 million development center for biotechnology pharmaceutical work by the late 1980s (Wysocki, 1987).

In the initial stages of biotechnology development, however, Japan lacked basic researchers with training in molecular genetics. As a consequence, the Japanese corporations, like their counterparts in the US, often formed contractual agreements with US biotechnology corporations. Japanese corporations lead all other foreign countries in the number of collaborations arranged with US biotechnology companies. Of the 165 collaborative ventures between US biotechnology companies and foreign corporations formed between 1981 and 1986, 71 (43 percent) were US/Japanese collaborations, the majority of which related to human health care (OTA, 1988). This pattern has continued with Japanese corporations forming 30 agreements with US biotechnology companies in the first half of 1987 (Klausner, 1987). Researchers at the North Carolina Biotechnology Center, however, observe that, while Japanese corporations will probably continue to form partnerships with foreign companies, this strategy appears secondary to building internal technological strength (Dibner and Lavich, 1987) and purchasing US companies (Cass, 1990).

Western European development of biotechnology, while quite diverse, has been equally impressive. As in Japan, the European industry has been concentrated in large, established corporations with large cash reserves and long-term research and development programs. Like the US and Japan, the European multinational corporations have gained access to basic research by forming collaborative agreements with US biotechnology firms. According to a recent OTA survey (OTA, 1988), from 1981 to 1986, European corporations were involved in 55 percent of the 165 collaborative ventures between US biotechnology companies and foreign corporations. In fact, while collaborations with US and Japanese firms have dropped or leveled off in the past three years, collaborations with European-based companies have been increasing.

European-based corporations have also established a significant presence in the United States. For example, at least ten major European corporations have major US operations that include research or manufacturing facilities. Among the motivating factors, particularly in Germany, are desires to escape what corporations perceive as strong public opposition and regulatory roadblocks to biotechnology in Europe. Badische Anilin und Soda Fabrik (BASF) recently announced plans to open a new biotechnology laboratory in Boston in 1990, and Bayer plans to open a facility for the production of recombinant factor VIII in Berkeley, California (Dickson, 1989). Moreover at least eight European companies formed US subsidiaries focusing on biotechnology, including multinational giants such as BASF, Bayer, and Boehringer-Mannheim from West Germany; Fisons and Imperial Chemical Industries from the United Kingdom; Elf Acquitaine from France; and Gist Brocades from the Netherlands (Dibner, 1986). In contrast, US biotechnology firms, attempting to form subsidiaries or affiliates in Europe, generally have not been successful − often selling their operations to European companies or scaling back their commitments. For example, Biogen liquidated its European holdings to Hoffman-La Roche and to Boots (Fowler et al., 1988), and Molecular Genetics reduced its investment in its Dutch agricultural affiliate, Mogen (*Bio/Technology*, 1989a).

This brief overview of foreign investments in biotechnology suggests the internationalization of commerce in biotechnology. Numerous government and industry studies have focused on the competitive positions of the US, Japan, and Western Europe in their efforts to commercialize new biotechnologies. However, the scenario that appears to be emerging is one in which no country will dominate. Instead, in all areas of the world where biotechnology is being commercialized, the large multinational corporations are playing an ever-increasing role. Consequently, the development and commercial control will likely be in the hands of the transnational corporations that transcend geographic boundaries and hold no allegiance to any nation.

Commercial Agricultural Biotechnology

The growth and development of biotechnology in the pharmaceutical and health fields has been paralleled by a similar, but as yet smaller, commercial development in agriculture. Biotechnology, however, is expected to play a major role in strengthening this important part of the global economy. As noted earlier, estimates of the market impact of biotechnology on agriculture by the year 2000 vary from a low of $12.6 billion to a high of $67 billion (Fowler et al., 1988). Estimated revenues from world seed sales are roughly $30 billion with the US share representing approximately one-fourth the world market. According to some analysts, after the turn of the century, approximately $12 billion of the estmated $30 billion world commercial seed market will contain contributions from biotechnology (Kidd, as quoted in Van Brunt and Klausner, 1987). Furthermore, packages including plants resistant to disease, pesticides, and herbicides could constitute a sizable portion of the $10 billion agricultural chemical market (OTA, 1988). In the area of food processing, one Japanese industry evaluation estimated the economic impact to be $17.2 billion in 2000, well above their $12.8 billion estimate for pharmaceuticals (Newell and Gordon, 1986). Others, not as optimistic, still anticipate real growth at a moderate 5 percent annually (OTA, 1988). A recent analysis of world agricultural markets (Ratafia and Purinton, 1988) predicted the probable years of commercialization for some genetically manipulated crop plants as follows: tomatoes, other vegetables, potatoes, and sugar cane, 1989; fruit, 1990; rapeseed, rice, and sunflower, 1991; and alfalfa, barley, corn, sorghum, soybeans, and wheat, 1992.

While most companies are pursuing applications of biotechnology in pharmaceuticals and diagnostics, a large number are involved in agricultural biotechnology. By 1984, 28 percent of the 219 US companies pursuing applications of biotechnology were involved in animal agriculture, and 24 percent were involved in plant agriculture (OTA, 1984). A worldwide survey in 1984 identified 322 companies from 16 countries involved in biotechnology, 114 of which were agriculturally related (*Genetic Engineering News*, 1985). In 1987, another worldwide survey identified 103 companies engaged in agricultural biotechnology (*Genetic Engineering News*, 1987). By early 1988, OTA identified 403 dedicated biotechnology companies in the US, 56 of which were involved in animal and plant agricultural research. Perhaps more important is the increase in investment in biotechnology by 70 major US corporations, 14 of which are committed to agriculture. Surveys of these agriculture-related firms reveal employment of more than 1,000 molecular biologists and annual investments in agricultural biotechnology research and development of over $200 million. Most of the agricultural biotechnology work continues to be conducted by US firms, followed by Europe, Latin America, Canada, and Japan (OTA, 1988).

The biological developments and the emergence of these commercial agricultural biotechnology firms have occurred at a time of major restructuring of the farm sector, concentration of the farm input and food processing industries, and increasing industrialization of food production. In the farm sector, an accelerating concentration has both reduced the number and increased the size of farms. In the US, from 1915 to 1945 the number of farms decreased by about 8 percent, while from 1945 to 1973 it decreased by 55 percent from 5.9 million to 2.8 million. By the late 1980s, the number was below 2 million. Market shares have also become increasingly concentrated. In 1970, the 4,000 superfarms (with annual sales of $500,000 or more) represented 0.13 percent of all farms, yet received 16.4 percent of the total net farm income. By 1981, there were 25,000 superfarms, representing 1 percent of all farms and receiving 66.3 percent of the total net farm income (USDA, 1982b). It is estimated that by the year 2000, 50,000 American farms will account for at least 75 percent of the nation's agricultural productivity (Klausner, 1986).

Today, less than 2 percent of the US population work in farming, while nearly 23 percent work in the food production system as a whole. Major segments of this food system have become input industries (seeds, agrichemicals, and machinery) and output industries (food processors, distributors, and retailers). In terms of their share of the food dollar, both industries are highly concentrated and are growing while the farming sector is declining. Overall, in 1979, 15 companies accounted for 60 percent of all farm inputs, 49 companies processed 68 percent of all the food, and 44 companies received 77 percent of all wholesale and retail food revenues (Vogeler, 1981).

Concentration is even greater for specific industries. By the 1970s, the top four tractor companies accounted for 83 percent of all tractor sales

(Vogeler, 1981). While 30 manufacturers in the US were engaged in pesticide development in the mid-1970s, there were only 12 by the late 1980s, and only six are projected to survive until the turn of the century. In the United Kingdom there were 60 manufacturers and formulators in the early 1980s, but only six were large enough to be important to the market. The pesticide industry worldwide is dominated by seven transnationals (Bayer, Ciba-Geigy, Du Pont, Hoechst, Imperial Chemical Industries, Monsanto, and Rhône-Poulenc), each with sales of $1 billion dollars or more, and together constituting 63 percent of global sales (Fowler et al., 1988). The seed industry has also been radically restructured. Estimates of the number of acquisitions of seed companies and other equity arrangements by international firms in the 1980s have ranged from 200 to more than 1,000. The top ten companies have almost 20 percent of the world's commercial seed market. Finally, the food processing industry is becoming increasingly concentrated. In 1987, the top 100 US food processors accounted for approximately 80 percent of all sales in the industry. Moreover, the top ten companies captured almost one-third of the market, up one-third from 1982. Mergers in Europe and the US have also dominated the industry. Over one-quarter of the 100 processors (in 1982) have been purchased or have merged. Recent activity among the top ten corporations includes the takeover of Nabisco by R. J. Reynolds, the purchase of Carnation by Nestle Foods, the acquisition of General Foods by Philip Morris, the purchase of Chesebrough-Pond's by Unilever, the merger of Con Agra with Swift, and the dismantling of Beatrice (Fowler et al., 1988).

This concentration, accompanied by horizontal and vertical integration across industrial sectors, reflects the mergers and acquisitions in the food processing industry, as traditionally nonfood industries dramatically expand their investments. This trend is also apparent in the input industries. Of the seven top pesticide corporations, five ranked among the world's 20 largest seed companies, with only Bayer and Du Pont having marginal seed interests. Moreover, of the ten top seed companies, eight (Ciba-Geigy, Imperial Chemical Industries, Lafarge, Pfizer, Sandoz, Shell, UpJohn, and Volvo) have significant interests in crop chemicals (Fowler et al., 1988).

The structure of the agribusiness and food processing industries is mirrored in the emerging agricultural biotechnology industry. While the large number of small, newly established, venture capital firms began as independent companies in the early 1980s and generated much of the early biotechnology research in agriculture, recent developments are characterized by consolidation, integration, and mergers. The agricultural biotechnology companies, like their counterparts in pharmaceuticals and chemicals, have recognized the need to merge and consolidate among themselves and to establish strategic relationships with the large agrichemical and seed corporations and multinational food processors. Similarly, the agricultural venture capital firms have also operated at a loss. They have found it increasingly difficult to raise additional capital, have encountered regulatory delays in conducting research and bureaucratic delays in obtaining patents, and have generally lacked the complete set of capabilities required to market a new product (i.e., research, development, formulation, scale-up, clinical and field testing, manufacturing, and marketing).

Some of the more active agricultural biotechnology firms in the late 1980s included Advanced Genetics Sciences, Agracetus, Agricultural Genetics, Allelix Biotechnica, Calgene Crop Genetics, DNAP, Ecogen International, Escogen, Molecular Genetics, Mycogen, Native Plants, Phytogen, Plant Genetics Incorporated (PGI), Plant Genetic Systems, and Sungene. Even these highly regarded and relatively well-funded firms have found it necessary to merge or establish collaborations with large multinational corporations (see table 1.1). For example, in 1988, DNAP – whose main technological strengths were cellular genetics and plant tissue culture – focusing on the development of new plant varieties and plant-based food products, acquired Advanced Genetics Sciences, gaining strength in molecular genetics, ice nucleation, and biological control technologies (Makulowich, 1988). In another major merger, Calgene acquired PGI in 1989. Calgene's molecular biology and plant biochemistry efforts complemented PGI's plant breeding and cell biology programs in alfalfa seed and potato. In addition, Calgene's operating seed company units provided a valuable means of distribution for PGI's product lines. Other recent acquisitions and mergers include Biotechnica International's 1989 purchase of the plant research business of Molecular Genetics (Bio/Technology, 1989b).

Finally, as noted earlier, the large multinational corporations specializing in oil, chemicals, food, and pharmaceuticals took the lead in agricultural biotechnology in the latter half of the 1980s (e.g., American Cyanamid, Campbell Soup, Ciba-Geigy, Du Pont, Eli Lilly, Hershey Foods, Lubrizof, Monsanto, Rhône-Poulenc, RJR Nabisco, Rohm and Haas, Shell, Sandoz, Standard Oil, and W. R. Grace). These corporations have begun to invest in their own in-house research and to establish research contracts, joint projects, licensing and marketing arrangements, equity positions, and control of ownership in the venture capital firms. The increasing number of collaborative relationships with the leading agricultural biotechnology companies can be seen in table 1.1. Most analysts predict that this pattern, in conjunction with increasing concentration and consolidation in the industry, will continue in the future. By the next century, a small number of highly diversified multinational corporations will likely control most of the food system.

Table 1.1   Selected major agricultural biotechnology companies: corporate collaborations and product focus

Company and year founded (location in parentheses)	Corporate collaborators (locations in parentheses)	Focus
Advanced Genetic Sciences, 1979 (CA)[a,b]	Du Pont (DE) Eastman Kodak (NY) Rohm and Haas (PA)	Tobacco "Ice minus" bacteria Bacterial soil additives
Agracetus, 1981 (WI)[c]	Cetus Corp. (CA) W. R. Grace (NY)	Crop improvements and microbial crop treatments in corn, soybeans, and cotton
Agricultural Genetics Co. Ltd, 1983 (UK)	Ciba–Geigy (Switzerland) Pioneer Hi Bred (IA)	Cereals, biopesticides Insect resistant corn
Applied Biotechnology, 1982 (MA)	Hoechst (Germany) Du Pont (DE) Mitsubishi (Japan)	Vaccine for chickens Pseudorabies and other porcine vaccines Equine vaccines
Biotechnica International, 1981 (MA) (Subsidiary: Biotechnica Agriculture Inc., 1987 (KS)[d]	RJR Nabisco (NC) UpJohn (MI) Monsanto (MO) W. R. Grace (NY)	Food products Silage inoculants Nitrogen-fixing crops Plants
Calgene, 1980 (CA)[e]	Bayer AG (Germany) Campbell South (NJ) Ciba-Geigy (Switzerland) DeKalb–Pfizer (IL) Kuraray (Japan) Nippon Steel (Japan) Procter and Gamble (OH) Rhône-Poulenc Agrochimie (France) Rohm and Haas (PA)	Herbicide resistance Processing tomatoes Disease resistant crops Herbicide resistant corn Agrichemicals Specialty oils for steelmaking Industrial and edible oils Herbicide resistant varieties Biopesticides
Crop Genetics International Corp., 1981 (MD)[f]	DeKalb–Pfizer (IL)	Bioinsecticides for corn
DNA Plant Technologies, 1981 (NJ)[a]	Campbell Soup (NJ) CIBA–Geigy (Switzerland) Du Pont (DE) Hershey Foods (PA)	Tomatoes Diagnostics Premium-quality vegetables Cocoa
Ecogen, 1983 (PA)[g]	American Cyanamid (NJ) EniChem (Italy) Monsanto (MO) Pioneer Hi Bred (IA)	Bioinsecticides Insect resistance Pesticide resistance Insect resistance in hybrid corn
Mycogen, 1982 (CA)	Japan Tobacco (Japan) Lubrizol (OH) Monsanto (MO) Shell Research Ltd (UK)	Bioherbicides Biotoxins Biopesticides Biopesticides, bioherbicides
NPI (Native Plants Inc.),[h] 1973 (UT)	DeKalb–Pfizer (IL) Sumitomo (Japan) W. R. Grace (NY)	Corn research New strains of tea and coffee Seasonings

Table 1.1 (Cont)

Company and year founded (location in parentheses)	Corporate collaborators (locations in parentheses)	Focus
Plant Genetics Inc., 1981 (CA)[e]	Kirin Brewery Co., Ltd, (Japan)	Synthetic seeds
	Merrill Lynch Tech. Ventures, L. P. (NY)	Genetic improvements in commercial potato varieties
	Monsanto (MO)	Potato research
	Weyerhaeuser (WA)	Woody products
Plant Genetic Systems, 1982 (Belgium)[b,g]	Clause Seed Co. (France)	Improved vegetables and potatoes
	Gist-Brocades and Amylum (Netherlands)	Protein engineering
	Japan Tobacco (Japan)	Tobacco biomass
	Hilleshög (Sweden)	Virus resistant sugarbeets
	Rohm and Haas (PA)	Development of biopesticides
Sungene, 1981 (CA)[i]	Lubrizol (OH)	Genetic engineering and tissue
	Mitsubishi (Japan)	culture to improve major agricultural crops (e.g., corn, sunflowers, wheat)

[a] Advanced Genetic Sciences merged with DNA Plant Technologies in 1988.
[b] Advanced Genetic Sciences owns 32% of Plant Genetic Systems.
[c] Joint venture of Cetus (49%) and W. R. Grace (51%).
[d] Biotechnica International purchased the plant business of Molecular Genetics in 1989.
[e] Calgene acquired Plant Genetics in 1989.
[f] Major stockholders included Merck's Co., John Hancock, Harvard University, and Elf Aquitaine.
[g] Collaboration between Ecogen and Plant Genetic Systems on insect resistance research.
[h] Company investors include Martin Marietta Corp., Sandoz, Elf Aquitaine, and British American Tobacco.
[i] In 1987, received US patents for regeneration of immature embryos in rice, sorghum, sunflower, wheat, barley, and corn.
Sources: Ratner, 1989; OTA, 1988; Klausner, 1989; Fowler et al., 1988; Bio/Technology, 1989b, 1990; Martin, 1990

Role of Patents

Finally, changes have been occurring in the patent laws and the court interpretations which have facilitated, if not stimulated, the emergence of biotechnology, created an enormous short-term administrative overload, and raised new debates about the impacts and ethical implications of patents for life forms. In 1970, Congress passed the Plant Variety Protection Act (PVPA) (7 USC § 2321 et seq.) which provided patent-like protection to new, distinct, uniform, and stable varieties of plants that reproduce sexually. This extended protection had already been accorded to new and distinct asexually propagated varieties other than tuber propagated plants by the 1930 Plant Patent Act (PPA) (35 USC §§ 161–164). More important, however, for the new biotechnologies were the landmark 1980 US Supreme Court decision, *Diamond v. Chakrabarty* (447 US 303), which provided complete patent protection for genetically engineered microorganisms, and the 1985 ruling by the US Board of Patent Appeals and Interferences, *Ex parte Hibberd* (227

USPQ (BNA) 443), which granted utility patents for novel plants. In contrast to the PVPA, these recent decisions provide exclusive patent protection for plants and microorganisms as long as they result from human ingenuity and intervention and meet the statutory criteria of novelty, nonobviousness, and utility (Buttel and Belsky, 1987; Lacy and Busch, 1989). Applications for utility patents for plants grew from 73 in 1986 to an estimated 400 in 1988. In addition, as of January 1988, there were nearly 7,000 biotechnology patent applications pending approval, a backlog likely to take five or six years to resolve (OTA, 1988; Naj, 1989). By the end of 1989, 15,000 biotechnology patents were pending (Schmeck, 1990).

In 1988, the Patent and Trademark Office granted Harvard University the first patent ever issued on animals for genetically altered mammals that carry certain oncogenes, making them susceptible to cancer. These genetically altered animals will be used for research to detect cancer-causing substances (Wheeler, 1988). An additional 44 animal patent applications were submitted by mid-1989 (OTA, 1989). OTA predicts that most of the early animal patenting activity will reflect interest in research animals, while later efforts will focus on agriculturally important species such as cattle, swine, sheep, poultry, and fish. OTA also notes that the largest economic sectors likely to be influenced by animal patents are the different markets for agricultural livestock and some segments of the pharmaceutical industry, although it is difficult to predict the effects of patents in the various diverse sectors of these industries (OTA, 1989).

Biotechnology has played three important roles with respect to the patenting of organisms, including plants and animals. First, it has served to increase interest in the issue of patents. Second, it has challenged the belief that plants are merely products of nature (Goldstein, 1986). Third, it has served to provide an apparent solution to the problems associated with the patenting of life forms. Let us consider each of these issues in turn.

The new biotechnologies have markedly increased the corporate interest in patenting because they open the possibility for much more rapid development of a wide variety of new products, such as plants selected for chemical tolerance, plants with special characters genetically inserted in them, and plants that interact with modified microorganisms to increase their efficiency (e.g., nitrogen-fixing bacteria). Because these types of modifications of life forms were not covered under the patent laws until recently, corporations have shown considerable interest in creating strong patent laws that are uniform across international boundaries to protect their investments.

The belief that life forms should be excluded from patent laws was also associated with the general principle that life is not under any circumstances a human creation. Biotechnology has challenged that principle by its apparent ability to create wholly new life forms. Plant breeding has always been limited to the genetic materials found in the parent plants; biotechnology opens the door to inserting any genetic material – in principle, even that from a bacterium or animal – into a plant. Thus, as evidenced by the term *genetic engineering*, biotechnology squarely puts plants into the same realm as machines, or at least it appears to do so. Indeed, in a recent international survey of industry opinion, virtually all the replies stated that there should be no difference in the application of patent laws to animate and inanimate matter (Beier et al., 1985).

In addition, biotechnology has been demonstrated to be of considerable use in testing patent claims. Specifically, by using electrophoretic techniques, it is possible to develop a picture of a plant's genetic map. No two varieties can be said to be different if their genetic maps are the same. Thus, in principle and in practice, such genetic mapping can serve to protect the profitability of an investment in plant research.

The impacts of the most recent changes in the patent laws are likely to be considerable, although there is significant debate over just what paths are likely to be taken. Some have argued that patenting will reduce the number of trade secrets because it will provide greater protection for invention in the field (Blumenthal et al., 1986a). For example, university scientists with industry grants are both more likely to have patents and to publish more than university scientists without such support. Fewer trade secrets, in turn, will mean a more rapid flow of scientific and technical information. On the other hand, others have argued that trade secrets will abound because patent laws require divulging information that might be of considerable benefit to competitors (Kenney et al., 1982; Beier et al., 1985). Nelkin (1984), discussing the overall issue of intellectual property, suggests that the problem may occur before patents are actually granted, for it is at that point in time that secrecy is truly essential. She notes a fundamental contradiction between the scientific norm of open communication and the need to keep such information secret until a patent is granted. Similarly, a recent report by the Council for Agricultural Science and Technology (CAST) (1985:29) notes that "the flow [of information] from companies to universities decreased [in the 1970s], perhaps in part as a result of the PVPA."

Evenson and Evenson (1983:210), making the case that patents will encourage smaller companies to enter into plant research, argue that "without patent protection, only oligopolistic firms have a strong incentive to engage in invention." Stallman and Schmid (1987), however, maintain that the evidence is weak at best that patents serve as a stimulus for research and development. Moreover, Blair (1972), in a classic work on economic concentration, argues that patents tend to reduce competition. Similarly, the recent developments in plant variety protection suggest it is the large chemical and pharmaceutical companies that have the necessary capital to do much of the research in this area.

Still another issue is that university scientists may organize their work around patentability rather than scientific importance (Yoxen, 1983). For example, Auerbach (1983:566) argues that "The existence of a protective patent system has the effect of encouraging scientists to proceed with inventions for which they can claim exclusive rights." While scientific and economic ends are not necessarily antithetical, it is clear that much of basic research is and will remain outside the patent system.

In addition to the scientific and economic issues raised by biotechnology patents, Congress continues to face three policy issues relevant to patents of life forms. First is the issue of whether the patenting of life forms, especially animals, should be permitted by the federal government. Some suggest a moratorium or complete prohibition while others seek specific statutes providing for such patents. Second is the question of whether the current statutory framework of intellectual property protection for plants is appropriate. Third is the issue of whether the current criteria for patents are adequate for biological material.

Embedded in these scientific, economic, and political issues, and perhaps of great importance, are the complex ethical problems associated with the patenting of life forms. The debates over the patents have focused on and highlighted many ethical and social issues that existed previously (e.g., animal rights, the distribution of wealth, the effect of technology on American agriculture, the release of novel organisms into the environment, genetic diversity, and soil erosion). The arguments have often focused on the consequences that could occur subsequent to the patenting of animals and plants (e.g., Who will benefit and how? Who will pay the price?), the inherent rights (e.g., rewarding innovation and entrepreneurship versus promoting a materialistic conception of life), and the impact on nature (e.g., Is nature enhanced, irreparably altered, potentially harmed, or made genetically more vulnerable?) Ultimately these are the fundamental issues of biotechnology that individuals and society must address.

Overview of the Book

This chapter has traced the scientific developments in biotechnology, the role that the federal government, state governments, universities, and industries are playing in its development, and the technical, social, and ethical issues associated with its emergence. Despite the infancy of the field, scientists, government officials, industry analysts, and the public perceive, predict, and envision a multibillion dollar market for engineered agricultural products in the foreseeable future.

Several impacts of the products of biotechnology are explored in this book. The processes by which these products are developed will affect the structure of the public and private agricultural research communities. The products themselves will affect the structure of agricultural research and the nature of the food system, not only in this country, but around the world as well.

We also examine the broad impacts of agricultural biotechnology on the organization and conduct of public and private sector agricultural research within this context and with these issues in mind. Although the analysis extends well beyond the case of plant genetics (i.e., plant breeding to biotechnology), we have focused on this research area for our examples and test cases. Our analysis draws on the relevant biological, social science, and philosophical literature as well as on in-depth interviews (one and one-half to three hours) with over 200 agricultural scientists and research administrators in both the public and private sectors.

The book will be of interest to a diverse audience, including agricultural scientists, social scientists, philosophers, and science policy analysts. Efforts have therefore been made to provide the necessary background information for all readers. This will facilitate their understanding of this complex interdisciplinary area.

Agricultural science is placed in the context of the relationship between science and society in chapter 2, and the nature of science and the underlying assumptions of agricultural science are examined. We present an analysis of the dominant philosophical and sociological views of science: positivism and the ideology of neutrality, and their implications in instrumentalism/technologism, scientism, and communication theories. These views are contrasted with those of science as nonneutral. We argue for the preferability of a version of the nonneutrality thesis.

Chapter 3 is a historical and sociological examination of traditional plant breeding and molecular biology and genetics. Here we contrast the methods and techniques of traditional plant breeding with those of biotechnology and compare the scientists who engage in each of these fields of science. The chapter concludes with a discussion of the implications of the move from a more holistic to a more atomized approach to science.

In chapters 4 and 5 we explore the general ideas described in the previous two chapters in the context of two specific cases. We examine, in detail, the changes occurring in research on an agronomic/field crop (wheat) and a vegetable/horticultural crop (tomatoes). We summarize the production, utilization, and structure of the industry surrounding each crop, other issues specific to each crop, a description of the scientific community, and the impact that biotechnology is having both on the science and the political economy of each crop.

The broader impacts that new agricultural biotechnologies may have on world markets and Third World nations are the subject of chapter 6. First, certain new technical directions of particular importance to Third World countries are discussed. Then, the consequences, in terms of changing world markets for agricultural commodities and restructuring terms of trade, are outlined. Finally, the difficulties that Third World nations will have in responding to this technical and economic challenge are described.

In addition to the biological, social, economic, and political issues surrounding the introduction of biotechnology to plant breeding, a number of important ethical issues also arise. Chapter 7 examines several of these issues including the privatization of public science, the decrease in public control over use of resources, the concentration of scientific talent in a limited number of organizations, the potential narrowing of the genetic base, and the implications the new agricultural biotechnologies have for the practice of science. We conclude the chapter with a discussion of the goals, role, and beneficiaries of public research, the proper role of the private sector, and, finally, the responsibilities that citizens, administrators, and scientists have with respect to science and society.

The last chapter addresses the implications of our findings for agricultural research policy. What is to be done, by whom, and for whom? After reviewing the current policy situation, we argue that the potential impacts of the new biotechnologies on agriculture, agricultural research, and plant breeding will not be simply the result of the inevitable progress of science. Instead, these impacts will emerge from numerous current decisions made by many actors in the agricultural and scientific communities. Our responsibility for ourselves and for our children demands that these decisions be made with care in the context of public debate and discourse.

2
Perspectives on Science and Society

From the beginning of civilization, agriculture has been foremost a testing of experience. Whether in the domestication of plants and animals, or in the development of strategies for dealing with a sometimes capricious natural environment, trial and error have predominated. Agriculture might even be called the first empirical science. Although prior to the refinement of the methods and tools of "modern science", and prior to the systematization and organization – indeed, rationalization – of the methods of agricultural production and of the understanding of the biological and environmental systems undergirding agricultural production, it could hardly be called scientific. Today, however, with the advent of modern scientific methods, and with the technologies that accompanied the historical development of those methods, agriculture is indeed a scientific enterprise. From the fundamental understanding of the nature of plants, animals, soils, environments, and institutions (especially, markets), to the application of those understandings to food production and processing, science predominates. Therefore, it is essential for our analysis of the implications of molecular biological research in the practice of agricultural science and agriculture that we address the questions of the nature and role of science.

In this chapter, we focus on prevailing philosophical and sociological views of the nature of science, with attention to the appropriateness of various models of science to the agricultural research enterprise. In particular, we show how a dominant view in the philosophy of science, often referred to as *positivism*, is inappropriate with respect to the nature of agricultural research, if not with respect to scientific research in general. Instead, we argue that a sociological perspective with, in particular, a view that regards agricultural science as part of the matrix of "economic" forces existing in a given society at a given time, is more revealing. Indeed, this perspective allows us to understand how it is that the nature of a scientific discipline – its methods, scope, agenda, and even its underlying philosophy – can be transformed in part by changes in the political arena and the marketplace. While a comprehensive review of the philosophy and sociology of science is not appropriate here, a brief, though admittedly cursory, treatment is essential.

Philosophical Assumptions in the Practice of Science

One of the dominant features of modern scientific practice is the tendency to think of science as a special human enterprise, governed by standards that are essentially different from other, ordinary approaches to knowledge and problem solving. These standards set scientific enquiry apart from other enterprises by virtue of being clear, generally well-formulated, rigorous, and fundamentally rational. By researching, testing, problem solving, and so on, in accord with these standards, individuals are engaging in scientific enquiry. Thus, this kind of enquiry differs from, for example, philosophical or musical 'discovery'. To the extent that a craft is understood and practiced according to these standards, the practitioner is a scientist.

Although modern science and its guiding standards developed from the earliest taxonomic enterprises of Aristotle and Ptolemy, as well as from the discoveries of Copernicus, Linnaeus, Bacon, and Descartes, it was not until the twentieth century that the standards became somewhat codified. While perhaps more representative of ideals than of the actual practice of scientific discovery and testing, the following characteristics are generally thought to be definitive of the methodological uniqueness of modern science:

1 Claims of knowledge of the world are justified by appeal to experience, experiment, or theory.
2 All cognitively meaningful statements are either true or false.
3 Truth or falsity is established either analytically (by consistency with the axioms of a system), or synthetically (by correspondence with facts of experience or experiment).
4 Causal laws are well-confirmed empirical hypotheses.
5 Theories are tested by their ability to explain known data and predict future occurrences.
6 Scientific fields are distinguishable from one another by their subject matter and not by their methods.
7 Science is defined by its methodology.
8 Judgments of aesthetic, political, personal, or moral value are irrelevant to scientific enquiry as *scientific* enquiry (Hollis and Nell, 1975).

These standards may be thought to form the philosophical basis for the specific methods and particular theoretical constructions which we now see operating throughout the sciences, including the agricultural and social sciences. As noted above, they can be thought of as ideals in the sense that the actual practice of research design, experimentation, and discovery in the laboratory would only approximate them. However, we would expect researchers to aspire to have their theories and hypotheses conform to them as much as possible. Insofar as these are ideal characterizations of science, they represent values that researchers come to accept in the process of education, socialization, and professionalization. Because they are uniquely scientific values, the injunction against allowing personal or moral values into science per se might still stand.

The collected philosophical standards listed above (there may be others as well) are frequently interpreted as representative of the logical positivist position in the philosophy of science. Historically, *logical positivism* was a school of the philosophy of science which sought, in accordance with standards 6 and 7, to "unify" scientific enquiry to provide a coherent, integrated understanding of nature, including human nature and the nature of societies (Joergensen, 1951). The collective efforts of philosophers such as Rudolf Carnap and Ernest Nagel, and the critical input from famous scientists such as Niels Bohr, Ernst Mach, and Albert Einstein helped to gain widespread acceptance for positivism. From the 1920s to the present, the tenets of positivism have been accepted, at least implicitly among scientists and explicitly among philosophers. Recently, however, positivism has come under attack, not so much for its intentions or for the seemingly unequivocal canons that we have listed above, but for the spirit of science that it seems to embody. In particular, many critiques focus on the apparently reductionist intent in the positivist approach to the ideal of the unification of science (e.g., Levins and Lewontin, 1985).

Reductionism

Reductionism is the term for any attempt to take integrated social or biological systems and explain them solely by the workings of smaller component parts. For example, according to a reductionistic program, societies are explained solely by the isolable, identifiable workings of individual members, individual human beings explained solely by the workings of their organs, organs explained by biological processes which in turn are explained by chemical processes. The ultimate reduction is to explain biological processes in molecular terms, which are in turn reducible to chemistry and ultimately to physics. Interestingly, most of the scientific proponents of positivism were physicists, and many of the specific theoretical formulations modeled after the notion of "ideal science" are physical laws and theories. More to the point, however, is the critics' notion that unifying science under the positivist canons logically *demands* a reductionist point of view. Indeed, if all sciences employ the same methods of discovery, hypothesis formation, testing, theory construction, and so forth, a reduction of wholes to increasingly smaller parts is more feasible than without a unity of methods.

This feasibility, however, does not imply that reductionism is a necessary consequence of positivism, at least not in theory. Presumably, each field in science, employing the same methods, could operate independently of other fields, with integration of the sciences a matter of seeing all the sciences from a God's eye view. Ideally, all fields of science might look the same, even if they were involved with different areas of nature. In short, while there is some sociological and historical force to the claim that the positivist ideal of a unified methodology of science would tend toward reductionism, logically at least, the critics of the positivist ideal are mistaken in their insistence that it requires a reductionist approach.

Value-Freedom

A more problematic aspect of the positivist view of science is its ideal of science as value-free, or at least free from nonscientific values. Science is a human enterprise and all human enterprises are marked by interests, desires, goals, motives, and ideals, so to claim that science is, in essence (i.e., ideally), practiced with no regard whatsoever for these desires is to posit a scientific ideal that is practically, if not logically, impossible (Feyerabend, 1975). Historically, we know that every scientific discovery occurred within some social context. These contexts are apparent in the conflicts that are endemic to science in the making, although they are obscured as science becomes "a body of literature" (Latour, 1987). Alternatively, individual scientists more often than not pursue a course of research because they think it is important. The problem with the positivist ideal is, the critics assert, that it fails to recognize these facts and hence is mistaken (if not internally inconsistent, since all facts should be considered in the formation of a theory, even a theory of science).

However, what critics fail to recognize is that as an ideal, value-freedom is neither inconsistent nor mistaken. It is, as contemporary neo-positivists have argued, another goal for scientific research to pursue (Quine, 1953). In other words, while no individual can be completely objective in his or her research activities, science is furthered by the attempts of the community of individuals to be objective, trying to let no values other than scientific values interfere with their work. Other scientific values might be appropriateness, fecundity of theories, simplicity of research design, honesty in reporting data, and the like. While from a sociological or historical perspective it can be argued that the positivist ideal fails, from the perspective of ideals for the practice of "pure science", the notion of value-freedom seems to stand (Ravetz, 1971).

Platonic Ideals

The major problem with the positivist position regarding what constitutes science per se, or pure science, is that it is not just an ideal; it is what philosophers refer to as a Platonic ideal. The question arises: what makes standards 1 through 8 (and others that might be added) definitive of science? Various answers have been provided for this question, from the claim that all practicing scientists simply know these canons constitute ideals for science, to the claim, 'that's what science really means'. None of the answers, however, are satisfactory. Indeed, a close look at the methods of various scientific disciplines – from botany to physics and from anthropology to political science – would suggest that approaches to subject matter differ greatly. Moreover, actual research methods, criteria for justification of conclusions, and kinds of predictions and explanations, differ greatly as well. In fact, science is not unified under the same kind of methods, nor, for that matter, would it seem necessary for it to be in order for knowledge about nature, including human nature, to be increased. Some positivists admit this and retreat to the claim that what is really essential is embodied in standards 1 through 3, and the value-freedom injunction. However, it should be noted that these statements are so open ended that they permit any number of methods to correspond to them, and may, with the addition of standard 8, be reduced to a single claim: science is concerned with establishing the truth or falsity of knowledge claims, without nonscientific goals or interests dictating or directing the course of the enquiry. Although there are, again, sociological problems with the second half of this statement, as a philosophical characterization of science, the statement is both innocuous as well as uninformative. It is another way of saying that "science is what science does" (Knorr-Cetina, 1981).

It is precisely this point that forces the philosophical analysis of the nature of science into historical and sociological realms. The answer to the question "What is (a) science?" essentially depends on an understanding of the workings of particular sciences, including both their methodological orientations and the social and value contexts in which they develop, grow, and mature. Though there are philosophical points to be made concerning the proper place of truth, falsity, different kinds of methods, and nonscientific values such as social utility or personal aggrandizement in the practice of a science, these points are dependent on an accurate understanding of how science exists and functions in a broader institutional context. Thus, it is to the sociology of science, and in particular, agricultural science, that we now turn.

Sociological Perspectives on Science and Society

The sociology of science (or rather, sciences) originally emerged from sociology proper, with historians and philosophers of science only relatively recently becoming dissatisfied with the conceptual models and Platonic ideals handed down from the logical positivists. In sociology, an attempt to understand scientific enquiry in the context of an ongoing social, political, and economic order developed from the work of the early "sociologists of knowledge", Max Scheler (1980), Karl Mannheim (1936), and later, members of the Frankfurt school of thought (e.g., Habermas, 1971). Philosophers and historians of science turned their attention to the social context of science in the 1950s and 1960s. However, it was Thomas Kuhn's (1970) seminal *The Structure of Scientific Revolutions* that set the context for further discussion in philosophical, sociological, and natural scientific circles about the nature of science as well as the relationship between science and society. Although a complete review of the sociology of science is beyond the scope of this book, a brief summary of some of the major positions should serve to clarify how science is affected by social and institutional actors as well as how it effects them.

Sociological perspectives on science fall into three basic categories which we label internal, external, and dialectical. Let us examine each of the three positions briefly.

Internal Community Views

Within this model, science is best understood as a community enterprise of generally like-minded individuals who share similar educational backgrounds, values, and approaches to problem solving, and whose values and approaches are communicated both internally and externally through institutionally established channels. Kuhn's aforementioned work, as well as the work of Lakatos (1970), Feyerabend (1978), and, with respect to the agricultural sciences, Busch and Lacy (1983), document the nature and dynamics of these more or less well-defined communities. Social psychological analyses are occasionally employed to explain how certain theories, laws, methods and conclusions are generated. Indeed, in this model, theories are developed and hypotheses are tested in a process of community response to, and provision for, problems perceived by both individuals within the community and "other forces". Other forces include markets, political direction, and funding agencies. However, it is the scientific community itself, somewhat in isolation from the greater society, that defines scientific methods, and indeed, the inherently "scientific" values to which scientists aspire. Thus, science becomes a sort of corporate organism, living in a non-scientific environment (Merton, 1973).

Externally Determined Views

For the most part, the classical sociologists of science regarded knowledge, methods, and values as being determined by external political or economic forces. Thus, Mannheim (1936) saw scientific knowledge as possible only at a certain stage in the historical development of a political–economic system (though he himself was unwilling to subject scientific knowledge to the scrutiny to which he subjected other forms of knowledge). Given a historically specific political economy, specific theories and explanatory schemes came to be fixed by social forces. According to this view, then, social values and social problems permeate to the core of even basic research, since all research arises unequivocally in response to a demand, a pressure, a political authority, or an economic market. Even values such as scientific objectivity or truth are definable in terms of the political–economic order. Interestingly, calls for a "politically correct science", for example, Lysenkoism, are only attempts to ensure that what must necessarily be an externally driven enterprise is driven in the right direction. Although the actual dynamics of how individuals fit into this sociological scheme are not always clear, what is clear in this model is that methods, theories, and other aspects of science are not derived autonomously by scientists; even discoveries are always explicable by the socioeconomic conditions of the time and place in which they occur (Brannigan, 1981).

Dynamic or Dialectical Views

In all probability, the majority of sociologists and sociologically oriented philosophers of science subscribe to some combination of the above. Science is neither entirely internally driven by its own community of practitioners, nor is it determined, at least mono-causally, by particular social or economic factors. Rather, there is some truth in the assertion that science – its methods, goals, research programs, and problem areas – is conditioned or limited by external forces, political or economic, while particular theories, hypotheses, explanations, predictions, as well as discoveries, occur autonomously. How a given discipline or field adapts to a changing political–social–economic context might also differ from how others adapt. Therefore, there are differences among the various authors who attempt to characterize the internal workings of a scientific field and its social "input" and "output", although there is fundamental agreement that sciences are never completely pure, nor are they completely sullied. Furthermore, there is fundamental agreement that the question of being pure or sullied goes to the very roots of the science, its scientific values, its methods, its theories, and its explanations. However, exponents of this "middle way" differ on the nature and extent of external influences on even these foundational philosophical matters, as well as over the autonomy of ordinary research activities (Ravetz, 1971).

Social science analyses of agricultural research are generally "middle way" views. Partly as a result of the recognition that the agricultural sciences were historically mission oriented and hence externally driven at least in part, and partly as a result of the nature of the agricultural sciences as sciences, the analyses of the relationship between the agricultural sciences and society or client groups have had to recognize degrees of interaction. However, differences exist in social perspectives on agricultural science regarding the extent of the influence of external forces on the logic, theory, and social structure of science. The remainder of this chapter is a discussion of the orthodox social science views of agricultural research held by sociologists and agricultural economists respectively (i.e., diffusion model and induced innovation model) and the heterodox approaches to agricultural science (e.g., client-driven science and science-driven research).

Orthodox Social Science Views of Agricultural Research

The Diffusion Model

According to the orthodox sociological model of agricultural research, borrowed from mechanical models of communications, scientists develop new technologies in their laboratories and fields, and educate extension agents who then disseminate them to farmers (figure 2.1). In addition, scientists train undergraduates who become farmers and extension agents as well as graduate students who become the next generation of scientists. Thus, an apparently complete picture of the system is provided in the three functions of teaching, research, and extension (Rogers, 1983).

This conventional model formed the basis for the diffusion research conducted through the land-grant schools from the 1930s onward. The

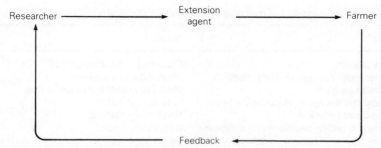

Figure 2.1   The conventional model of the research and extension process.

practical goals of this research were to increase the rate and complete-ness of adoption of agricultural innovations. Feedback, while added later to the model, consisted largely of the farmer informing the resear-cher whether or not the innovation had been adopted. Thus, although the communication was theoretically two-way, it was strongly biased in favor of the researcher. The researcher was assumed to have the right question and the right answer to the problem, and failure to adopt the innovations was best understood as the farmer's stubbornness or ignor-ance.

What the conventional model failed to do was to distinguish between scientific or narrow means–end rationality and everyday problem-solving rationality. Scientific rationality is the approach generally used in the process of "doing" science. Consider, for example, table 2.1. In this situation, the *end* of the research is to increase agricultural produc-tivity. Various *means*, including those listed in the table, are available for achieving that end. In general, the researcher will choose the means that are most suitable to his or her disciplinary background. While the choice of means is discussed at length, the end is taken as a given, even though it is by no means a final one (it is itself a means to an end).

A central feature of scientific rationality is that it is instrumental (Idhe, 1979; Busch, 1984). This instrumental character of science is

Table 2.1   An illustration of scientific rationality

End	Means
Increase agricultural productivity	Breed more productive plants
	Control pests
	Reduce Disease
	Increase soil nutrients
	Modify cultural practices
	Introduce irrigation
	Develop more efficient machinery

Table 2.2   An illustration of everyday problem-solving rationality

Ends	Means
Make a profit	Grow high productivity crops
Demonstrate success to one's neighbor	Mechanize production
Minimize heavy toil	Minimize cost of purchased inputs
Spread one's work evenly over the year	Increase farm size
Minimize soil erosion	Find nearby markets
Maintain a certain degree of independence	Construct terraces
Insure a minimal harvest even in poor	Select varieties over hybrids
years	Select seeds primarily for yield
Minimize genetic erosion	consistency
Maximize genetic diversity	Select appropriate intercropping and crop/
	livestock balance

manifested in several ways. First, science involves the use of instruments (e.g., spectrometers, pH meters, microscopes, scales) to construct knowledge. We need only walk through a modern scientific laboratory to note the profound importance that instruments have for science. Second, science is instrumental in that it is concerned with the choice of means and not of ends. While individual scientists may employ symbolic, analogical, or even literary reasoning in their work (Knorr-Cetina, 1981), the end toward which they strive is not called into question. To return to the example presented in table 2.1, the decision to increase agricultural productivity is the proper subject of philosophy or politics but not of science.

Now consider what we shall call everyday problem-solving rationality. In everyday situations, both the ends and the means may be more varied and their interactions more complex, as can be seen in table 2.2. It is immediately apparent that the choice of both the ends and means is much more open-ended than in the case of scientific rationality. Moreover, not all the ends can be maximized at once. Thus, tradeoffs, rankings, or some integration of them will be essential (Aiken, 1986). The same will be true of the means: the degree to which a given means is employed will be based on the relative importance attached to attaining various ends. While an economist might be able to calculate an optimal solution given a particular set of weights for the ends, by virtue of the complexity of the decisions and the time pressures on the farmer, a simple rational calculation of an optimal solution may not be the best path to pursue.[1] It is more likely that a farmer will decide what to do on

---

[1] The recent development of computer technology for use by farmers in decision making makes the ideal of rational calculation more realizable than before.

the basis of what appears most appropriate given a range of constraints. Science and technology may only aid in attaining some of the desired ends.

In rational scientific practice, scientists not only control for those factors that might directly intervene in an experiment, but they also control or even ignore other considerations such as the larger political and socioeconomic environment. They must limit their research to the service of one or two relatively well-defined ends. Farmers rarely attain or even desire this narrowing of options. This difference between scientific and ordinary problem-solving rationality has profound effects on communication.

Problems with the diffusion model are clearest against the background of their application in technology transfer to Third World nations, although the problems are apparent in the Western World as well. Important assumptions incorporated into the diffusion model are discussed at length by Busch (1978) and include the following.

Ontological monism   Diffusion theory assumes that there is a single, objective social and physical world in which we all live. Ethnographic research in non-Western societies challenges this assumption by showing that knowledge of many aspects of a culture is only attained through participating in it; those who participate, in some respects, see the world differently. In a very real sense, people in radically divergent cultures live in different worlds. A similar case might be made for the differences between scientists and farmers (see Latour, 1987).

Objectivity of technical language   Diffusion theory gives little or no attention to the problem of differences in the purpose of language use. Translation of scientific language to lay language or from one ordinary language to another is seen as a relatively straightforward task. In contrast, many philosophers and sociologists have shown that technical language is not easily translatable into ordinary language. For example, the categories of scientific botany and those of the ethnobotany of an African society reflect the different concerns of the two groups. Therefore, neither is objective nor necessarily true or false.

Communication as monologue   This aspect of the diffusion perspective follows from the two points just discussed and is apparent in figure 2.1. There is no meaningful feedback to direct the research or to inform researchers that a particular innovation is unsuitable or in need of modification. If researchers fail to see perspectives of people from other cultures or the relativity of their own categories and language, then communication between scientists and laypersons is stifled.

Tradition versus modernity From within the diffusion perspective, it is always "we" moderns against "those" traditionals. This perspective, first developed in the self-conscious modernism of Descartes ([1637] 1956), denigrates tradition as inherently erroneous. Tragically, it is often manifested most extremely by members of Third World societies who have received scientific training in the West. Ironically, this position ignores the important, perhaps pivotal, role that tradition plays in science. It is not accidental that scientific journal articles begin with a review of the literature; like all traditions, those of science have withstood the test of time. They have been declared to contain the truth, verified by a (fallible) scientific community. The disregard for tradition has often led scientists to jettison the experience that Third World people have of their particular agroclimatic zones.

Separateness The diffusion approach also treats science quite differently from the rest of society. Science is an institution from which finished innovations emerge much as they did in Bacon's mythical House of Salomon. They are then either adopted by a willing audience or rejected by an unwilling audience. As Latour (1987) has suggested, in this view science is separated from the rest of society and becomes an almost irresistible force.

In sum, the diffusion approach has been limited by a lack of understanding of the difference between scientific and everyday rationality, as well as by a failure to recognize its own underlying assumptions irrespective of whether everyday rationality is manifested by a Third World farmer or a Western one.

The Induced Innovation Model

The induced innovation hypothesis was initially mentioned in passing by Hicks (1932), but has only recently been developed into an articulated theoretical perspective (e.g., Binswanger and Ruttan, 1978; Ruttan, 1982). The argument put forth by its proponents is fairly straightforward: agricultural scientific and technical innovations are said to be developed in response to relative factor scarcities. Thus, in nations where labor is scarce, innovations will tend to be laborsaving, while in nations where land is scarce, innovations will tend to be landsaving. The United States and Japan, respectively, are frequently used as examples of this marked difference in research trajectories. In contrast, Elster (1983) notes, some argue that entrepreneurs, including farmers, will accept any cost-reducing innovation, not merely those that reduce the cost of the scarcest factor.

One key assumption of the model is that returns to research will accrue to the firm conducting the research and development. In most countries this occurs as a result of patent laws. Patents grant exclusive license to market a given innovation to a given firm for a given period of time. However, other government policies may also change the rate and direction of technical change. Binswanger (1978:18) explains: "If the rate of technical change in an industry is responsive to the price of that industry's output, then policies that alter prices of the output of one sector of the economy will affect the rate of technical change of that sector and of the sectors that produce substitutes." Such policies might include subsidies, high import or export taxes, regulation of the use of certain technologies, and so on. For example, Sanders and Ruttan (1978:277) note that "government resources used to subsidize tractors [in Brazil] and to subsidize domestic industrial capacity to produce tractors must be directed from other uses such as the creation of yield-increasing biological technology." These same policies also served to shift sugar production from small farms in the Brazilian northeast to large farms in the south.

Since much agricultural research has not traditionally yielded patentable products, there has been little incentive for the private sector to invest in it. Hence, early American agricultural research focused almost exclusively on machinery (Wik, 1966). Only with the development of USDA research capabilities and the passage of the Hatch Act in 1887 establishing the SAES did biological, chemical, and, later, social science research in agriculture become established. This same division of labor between the public and private sectors in agricultural research is apparent in virtually all the market economies of the world. However, public sector research is outside the market.

> [When] research is publicly funded, the research resource allocation process becomes as imperfect as any public allocation mechanism. The latent demand for technical change must be filtered through political institutions, and the outcome depends heavily on the political influence of various groups whose income position stands to be affected by the technical change. Efficiency considerations are not the only criteria by which choices will be made. (Binswanger, 1978:15)

Nevertheless, Ruttan (1980) argues that, despite the bureaucratic nature of public research, public sector research institutions are quite efficient. He suggests that this may be due to the highly competitive character of farming. In addition, a review of studies of returns to public sector research in a large number of countries, using various estimation procedures, shows consistently high returns to research investment (Ruttan, 1982). Moreover, Trigo and Pineiro (1982) note that when the state fails to provide effective agricultural research institutions, powerful interests in the private sector will attempt to provide those services themselves.

Ruttan (1982) has extended the induced innovation approach to institutional innovations. He argues that effective institutions innovate in response to changing environments. Such institutional changes include changes related to the conditions of land ownership, development of organizational structures that effectively produce research, and incorporation of social science research that identifies bottlenecks and weak links in the agricultural production system and in the research institutions themselves. A particularly interesting example of this latter point was a study of the economics of weed control conducted at the International Crops Research Institute for the Semi-Arid Tropics (ICRISAT). The study revealed that chemical control was much more expensive than hand and animal control in the nations ICRISAT serves and, in addition, displaced landless laborers. As a result, chemical control of weeds was de-emphasized at ICRISAT (Binswanger and Ryan, 1979).

While the induced innovation perspective represents a significant step forward in our understanding of agricultural research and our ability to improve its effectiveness, it has not been without its critics. For example, Pineiro et al. (1979) have argued that, in Latin America, inducement mechanisms described by Binswanger and Ruttan (1978) have not resulted in innovation. They note that "different social groups, and particularly those directly related to agricultural production, will have different attitudes towards technology depending on their expectations of the effects of the technology and their capacity to appropriate the potential economic benefits derived from its utilization" (Pineiro et al., 1979:172). They also note that the state intervenes in creating both the supply of and demand for agricultural innovations. "A central point is that demand and supply are interdependent through the role played by the state in the determination of model components which affect both sides [of the equation]" (Pineiro et al., 1979:174). Since supply and demand for research can in no sense be considered independent, the usual assumptions of mainstream economics can no longer be considered valid.

DeJanvry and LeVeen (1983:27) pursue the consequences of this line of reasoning: "Technical change conditions the social control of the means of production, the organization of the labor process, the social division of labor, and the social appropriation of the surplus. As such it is a powerful instrument of social change or stasis." Therefore, research directions and appropriations are determined in large part through social conflict rather than by purely technical means (e.g., optimizing agricultural incomes or crop yields). They note that rising labor costs in California were the result of changing social relations rather than any scarcity of labor. In addition, the California experiment station undertook research on mechanization long before any problem was made manifest (Friedland et al., 1981).

A related criticism concerns the international dimensions of agricultural research. Much research in the Third World is as capital intensive as that conducted in the United States. In addition, it has often centered on products of interest to the developed nations and has employed modes of production more appropriate to developed nations than to those of the Third World (Trigo and Pineiro, 1983). In short, not only national groups, but international groups as well, participate in the research resource allocation process. Certain kinds of indigenous research, even in the private sector, may be suppressed or discouraged by more powerful international interests.

A final point concerns the implicit assumption of a democratic society embedded in the induced innovation perspective. Public support for research is seen as the result of an open policy-making process that, although it may favor some interests over others, nevertheless allows all to be heard. In many Third World nations, such an open process simply does not exist. Steinberg et al. (1981) and, especially, Burmeister (1987) suggest this to be particularly true of Korea. Burmeister observes that directed innovation might be a more apt term to describe the situation in that nation.

In sum, the induced innovation approach offers a substantial number of insights over and above those of either diffusion theory or mainstream economics. Nevertheless, it tends to adopt a tacit consensus perspective. Indeed, if diffusionists see science as largely autonomous, proponents of the induced innovation perspective assume that scientific institutions are so in tune with society that they automatically modify their outputs to conform to societal change. Recent developments in the sociology of science, however, complement the induced innovation perspective well. Two schools may be defined, focusing on the role of clients in the broadest sense in creating a demand for science, and focusing on the role of scientists and administrators in creating a supply.

Heterodox Approaches to Agricultural Science

Client-driven Science

As discussed earlier, many sociologists traditionally have conceived of science as a largely autonomous activity conducted by scientists with little concern for the larger social world. Knorr-Cetina (1981), in *The Manufacture of Knowledge*, has challenged that view. After observing a food scientist at work for a year in a university laboratory, she coined the term *transcientific fields*, meaning not only those who work within a given research group but also nonscientists who have an interest in the outcome of the research. She notes that scientific journal articles are written to serve a special purpose: "Scientific papers are not designed to promote an understanding of alternatives, but to foster the impression that what *has been done* is all that *could be done*" (p. 42). This, however, conceals the complex negotiations over just what will be done (Busch, 1980). Moreover, it is misleading to consider the influence of nonscientists on research problem choice as an external influence: "To refer to research problems as an external input ignores the fact that the process of defining a problem penetrates to the core of research production through the negotiations of its implications and operationalizations" (Knorr-Cetina, 1981:88). In short, Knorr-Cetina argues that clients cannot be seen as outsiders but form an integral part of each transcientific field.

Wolf Schafer (1983) and his colleagues take a different path but arrive at a similar conclusion, arguing that science has become a social resource which can be aimed at the solution of various social problems. Bohme et al. (1983) argue that scientific fields may pass through three phases: first, the exploratory or preparadigmatic phase where discovery rather than theory is the rule; second, the paradigmatic phase in which the problems of research are determined by theory (e.g., Kuhn, 1970); third the postparadigmatic phase, during which theoretical issues are finalized. The methods and exemplars of the research group are well defined and subject to little debate. Moreover, there is no compelling theoretical reason for pursuing one research trajectory as opposed to another. At this point the practical concerns of the larger social world begin to take a central role in guiding research.

To illustrate their perspective, Krohn and Schafer (1983) turn to agricultural chemistry, noting that agricultural chemistry was developed as a separate field by Liebig in response to Malthus. Liebig reasoned that the only way out of the Malthusian dilemma was to increase agricultural productivity by developing the scientific specialty that came to be known as agricultural chemistry. Liebig's agricultural chemistry "... emerged as a science not only to explain the processes of agriculture, but also to *shape* them in accordance with human purposes" (Krohn and Schafer, 1983:29). In short, for Liebig, agricultural chemistry was at once a science of natural cycles and a technology that could be used to alter those cycles for human purpose.

It takes little extrapolation from this approach to realize that agricultural research itself consists largely of "finalized sciences". Moreover, the commitment to application, the mission orientation of agricultural research, ensures that clients are central in directing it. Hence, while the basic principles of plant and animal physiology and pathology were worked out many years ago, research is conducted to develop special theories that explain the physiology of corn or cows, because these organisms (commodities) are of concern to client groups.[2]

A third perspective on the role of clients is offered by Busch and Lacy (1983), who attempt to answer the question: How are research problems formulated in agricultural research? To arrive at an answer, they reviewed the official literature of the SAESs and USDA from when they were founded to the present, conducted in-depth interviews with scientists in several locations, administered a mail survey (to which 1,431 scientists responded, a 76 percent response rate), and reviewed the recent technical literature for state-of-the-discipline statements.

When scientists were presented with a list of 21 criteria for problem choice they ranked "feedback from extension" as twentieth. In contrast, "demands raised by clientele" ranked thirteenth and "client needs as assessed by you" ranked seventh. In addition, 36 percent of the scientists rated "client or potential user" as an influential person in their choice of research problems. In contrast, colleagues were seen as influential by only 20 percent of the respondents. In short, extension had little to do with problem choice, but clients did, even though client demands were not simply responded to in a passive way. To the contrary, scientists were more likely to make their own assessments of relevance. The words of one respondent captured this notion.

> Researchers in agricultural economics (as in most disciplines, I suppose) have difficulty in determining what research would be most useful. They prefer to research those areas in which a lot of people would appreciate getting the results. The public and their other clients, however, do not communicate their needs well so the researcher has to decide on his own what is important. (Quoted in Busch and Lacy, 1983:47)

In short, Knorr-Cetina, Schafer, and Busch and Lacy emphasize that science is conducted in response to client demands – demands expressed not through the market mechanism but through negotiation, persuasion, and coercion. Successful science, it would appear, must respond to those demands rather than go off on interesting theoretical paths for which there is no demand. But if there is a demand side to science, then presumably there must be a supply side as well.

---

[2] This has often been carried to extremes. Researchers in agriculture have often felt it necessary to confine their activities to crop plants while those in the basic sciences have studiously avoided them (Levins, 1973).

Science-driven Research

Bruno Latour (1984, 1986, 1987) proceeds by asking whether science can be separated from politics. He answers this rhetorical question with a resounding no. Louis Pasteur is often viewed as one of the great scientific geniuses of the nineteenth century. He is often described as a lone genius who made his contributions to science and medicine through dedication and hard work. While not denying Pasteur's genius, Latour demonstrates that he was also a great organizer: "Pasteur, from the start of his career, was an expert at fostering interest groups and persuading their members that their interests were inseparable from his own" (Latour, 1983:150). Latour shows at great length how the Pasteurians positioned themselves between the social world and the world of microbes. Only the Pasteurians had access to this microworld, and only they seemed to be able to reproduce that microworld in the laboratory. The hygienists could produce statistical relations between diseases and certain physical phenomena (e.g., raw sewage, standing water, polluted air), but only the Pasteurians could recreate that relationship in the laboratory. Moreover, every time that Pasteur encountered an "applied" problem, he turned it into a fundamental one to be resolved by disciplinary methods.

The Pasteurian laboratory, however, has certain very important characteristics: "The laboratory positions itself precisely so as to reproduce inside its walls an event that seems to be happening only outside . . . and then to extend outside to all farms what seems to be happening only inside laboratories" (Latour, 1983:154). Thus, Pasteur first brought animals into the lab and created an epidemic; he then literally brought his laboratory into the field and prevented anthrax by convincing French farmers to vaccinate their sheep and keep their barns clean — in other words, to make their barns, as much as possible, resemble his laboratory.

In short, Latour argues that Pasteur's success was due not only to his genius as a scientist but also to his ability to organize various interests to transform the world. It takes only a few moments' reflection to realize that the same may be said for both the successes and failures of the Green Revolution of the 1960s. It was the scientists and administrators, first in the Rockefeller and Ford Foundations and later in the International Agricultural Research Centers, who identified the need for high-yielding varieties — not Third World smallholders. Then they set about developing high-yielding varieties on experiment station fields and in their laboratories. Yield response under optimal levels of fertilizer, water, light, temperature, and so on was the goal of much Green Revolution research. The high-yielding varieties were diffused by convincing (some) farmers to reorganize their fields — introduce irrigation, fertilizers, pesticides, and various new cultural practices – so that the fields more closely resembled the experimental fields of the researchers. Only those farmers who had the wherewithal (e.g., capital, access to credit, and inputs) to replicate the research plots on their farms were able to benefit from the Green Revolution bonanza.

Moreover, the Green Revolution's effects were not merely technical. Once the new varieties were in place and the old ones abandoned, changes of an irreversible sort took place. Old patterns of landlord-tenant relations no longer existed. The class of landless laborers increased in size. Some farmers became wealthy. Bullock carts were abandoned and replaced with tractors. As Winner (1986:48) put it, in a somewhat different context: "It is not merely that useful devices and techniques of earlier periods have been rendered extinct, but also that patterns of social existence and individual experience that employed these tools have vanished as living realities."

The examples of Pasteur and the Green Revolution clearly show that scientists can and do package and sell their products and processes to the larger public. Moreover, they often do this aggressively. Thus, not only clients but also scientists themselves contribute to both the scientific agenda-setting process and the determination of the products that emerge from scientific research. In so doing, they restructure both the social and technical aspects of life – indeed, they blur the distinction between the social and the technical. What remains is to integrate the sociology of science perspectives with the induced innovation approach. That is the subject of the following section.

Synthesis: Supply and Demand in Science

Figure 2.2[3] attempts to synthesize (in simplified form) the various perspectives described above. The various clients/users impinge on the choice of research problems that the researcher addresses (Busch and Lacy, 1983). They may do this through funding mechanisms, by lobbying funding organizations, by direct pressure on the research organization, or by administrative decree. Moreover, each of the client groups will want different things from the researchers. For example, farmers may desire lower production costs, agribusinesses may desire greater use of fertilizers, and administrators may desire annual progress reports or large numbers of articles in scientific journals. In contrast, government agencies may desire new seeds that can be multiplied, scientific disciplines may desire major breakthroughs, and legislators may desire a reduction in urban food prices. Thus, the variety of groups

---

[3] Much more detailed analysis of the structure of contemporary US agricultural research can be found in Busch (1980) and Busch and Lacy (1983).

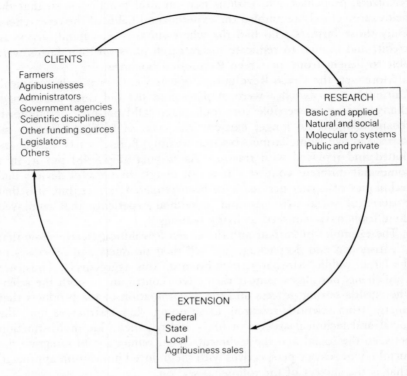

CLIENTS
Farmers
Agribusinesses
Administrators
Government agencies
Scientific disciplines
Other funding sources
Legislators
Others

RESEARCH
Basic and applied
Natural and social
Molecular to systems
Public and private

EXTENSION
Federal
State
Local
Agribusiness sales

Figure 2.2   An alternative view of the research and extension process.

that create a demand for agricultural research is substantial. However, unlike the demand for a commodity such as wheat, which is cumulative and quantitative, the demand for research is, in reality, a set of partially competing and conflicting demands. The problems finally selected for research arise out of negotiation, persuasion, and coercion involving the full range of clients and the researcher(s) (Busch, 1980).

In addition, the researcher substantially controls the supply of research because he or she is one of a very small group – in small organizations, perhaps the only one – who fully comprehends the ways in which research can be brought to bear on client demands. Moreover, research cannot be produced on an assembly line like automobiles, but instead requires arcane skills and instrumentation. The researcher retains substantial latitude in defining the research even in the most hierarchical research organizations. The practice of doing one thing and calling it something else is so commonplace that it has acquired a clear designation: bootlegging (Greenberg, 1966). Moreover, a competent researcher is likely to have, as a result of training and background, many independent ideas which enter into the negotiation, persuasion, and coercion preceding and during research. Finally, if the researcher has been a member of the research organization for some time, he or she will understand the reward system. The research program selected will reflect those rewards, whether they consist of filing annual reports on time, publishing many journal articles, or working in farmers' fields. The outcome of this complex process will, necessarily, set limits on the range of products that are provided to extension agents for diffusion or to agribusiness for product development.[4]

In sum, innovations are induced, in part, by the relative scarcity of the factors of production. However, the relative scarcity of the factors of production is only one of many considerations that enter into the public research decision-making process. Indeed, as Latour (1987) points out, a notion such as relative factor scarcity is itself an artefact of an efficient system for collecting economic data and analyzing them. Of necessity, many other considerations, some of which are only remotely related to commodity production, impinge on the research process. This is true even of large private organizations with research and development laboratories, where conflicting pressures from sales and production staff, as well as researchers' own interests, must be taken into account (though undoubtedly the bottom line plays an important role).

If researchers do their homework, then the products created are those that were demanded by clients in the first place. The diffusion process largely involves making those products and their uses known to the clients that requested them.

Of course, all members of each client group are not alike. Farmers may be wealthy or poor, may grow different crops and livestock, may or may not hire labor, and may have very different interests. Similarly, as noted above, legislators may be more interested in keeping urban food prices low than in augmenting farmers' incomes. The diffusion literature has often noted the tendency for better capitalized, better educated, higher status farmers to adopt innovations more rapidly than their neighbors (Rogers and Shoemaker, 1971). This is not due to any innate propensity to adopt on the part of these individuals; it is, rather, because they are more articulate and have greater access to the research system itself. Ruttan (1980:540) explains why:

> Under competitive market conditions the early adopters of the new technology in the agricultural sector tend to gain while the late adopters are forced by the product market "treadmill" to adopt the new technology in order to avoid even greater losses than if they retained the old technology. One effect of the tread-

---

[4] This is necessarily the case for any research system as a result of limits on financial and other resources.

mill phenomenon is ... to limit the economic motivation for [farmer] support of agricultural research to a relatively small population of early adopters of new technology. The early adopters also tend to be the most influential and politically articulate farmers.

In short, various clients will have differential access to researchers depending on their wealth, power, status, class, and even their ability to articulate their demands to researchers. Obviously, in societies in which income, wealth, class, and status differences are already pronounced, the problem of differential access will be proportionately magnified.

The recent move to farming systems research (FSR) represents an attempt to overcome the problem of differential access.[5] By collecting information about problems directly from smallholders, the probability that research will directly serve their needs is enhanced. However, we should not look to farming systems research to resolve the technology treadmill problem. That problem is endemic to competitive markets. Simply put, the early adopters capture most of the gains from adoption.[6] If they use those gains to increase the size of their farm operations, as is often the case, then they are in an even better position to be early adopters of the next innovation. If this process continues for a long time, then the distribution of farms by size becomes bimodal irrespective of the scale bias of the innovations.[7] A few large farms grow most of the marketed agricultural produce while a larger number of small farms produce mostly for home consumption. The medium-size farms disappear, their owners migrating to the city in search of more lucrative employment. A recent study, examining this problem in the United States from 1915 to 1973, concluded that public agricultural research significantly increased farm size during that period independently of other contributing factors (e.g., debt, taxes, unemployment) (Busch et al., 1984).

In summary, this synthesis adds several often-neglected dimensions in an attempt to understand agricultural research systems. First, the importance of client demand and researcher supply is clearly stated. Second, the supply of and demand for research are not merely market functions but must, of necessity, enter into the larger political sphere.

[5] Key approaches include Hildebrand (1980–1) and Norman (1978). For a review and critique see Oasa and Swanson (1986).

[6] The very first adopters of a new technology may actually lose from adopting it. Often, at this early stage in its development, the technology is not yet fully adapted for the use to which it will be put. Further refinements are necessary before its use becomes profitable to a wide range of potential adopters (Rosenberg, 1982).

[7] Note that this bimodal distribution is primarily true of farming. This is largely due to the fact that farm produce can be consumed by the producer. In contrast, the treadmill tends to eliminate all but the largest producers in other industries.

Third, some clients have more access to the research system than others. Fourth, diffusion largely provides certain clients with the innovations that they initially requested. And, finally, the necessary linkage between agricultural research policy and agricultural research is clarified.

These linkages – between technical changes, markets, and the social and political sphere – form the basis for our analysis of the plant improvement process. Since the specifics of these relations vary by commodity, we have focused on two that are of particular importance: wheat and tomatoes. However, to avoid mechanistically applying the framework developed above, we need to look more deeply at the historical, social, and technical processes involved in plant genetic research. This is the subject to which we turn next.

3
From Plant Breeding to Biotechnology

Plant genetic research using traditional plant breeding and the new biotechnologies differ in a number of significant ways. Most apparent are the differences in the focus and technical details of the two disciplines, the topics scientists study, and the level of analysis. Variations also exist in how the two groups of scientists view and approach their work. The different research orientations of the scientists stem partly from the different histories of the two disciplines, as well as from differences associated with the subject matter, and raise some larger issues in the philosophy of science (e.g., the debate surrounding reductionism). In addition, the two fields differ in terms of the "big" interdisciplinary science of molecular genetics and biotechnology and the little individualistic science of classical genetics and plant breeding. Finally, there is variation in the social background of the scientists themselves, and in the status of their respective disciplines within the scientific community. At the same time, the two fields share an emphasis on the control of nature and crop improvement.

This chapter should not be viewed as a summary of the history of plant breeding and plant molecular genetics. Instead, it is an attempt to use the theoretical perspective developed in chapter 2 to shed light on these fields, as well as to use the empirical evidence to qualify the theory. In short, we argue that both molecular genetics and plant breeding have been characterized by a supply and a demand. Moreover, the supply and the demand for each field have been mediated through negotiation, persuasion, and coercion. The social and historical roots, similarities, and differences between traditional plant breeding, and molecular genetics and biotechnology are the focus of this chapter. First, we briefly examine the history of plant breeding, an activity that predates modern science, and follow this with a description of the traditional breeding techniques – what might be called the old biotechnologies. Next we review the historical development of molecular biology and genetics. This is followed by an analysis of the new biotechnologies (i.e., cell and tissue culture technologies and genetic transfer technologies). Then, we briefly examine the scientists themselves in terms of their background, their education, and the relative status and prestige of their disciplines. We conclude with a brief comparison of the old biotechnologies with the emerging biotechnologies.

History of Plant Breeding

Until about 200 years ago, the "science" of plant breeding did not exist. Moreover, it was not until after 1900, with the acceptance of modern genetics, that plant breeders gained new knowledge of the processes by which certain characteristics were passed from one generation to the next. Farmers and peasants were the forerunners of today's plant breeders; they grew their own crops, selected the "best" plants (according to their desires) and used seed from those plants for the next crop. With wheat, for example, one would save the largest seeds to be planted the next season. For perhaps thousands of years, humans have selected wheat in this manner. In classical Greek times, a mass selection technique was well known. Columella (1911) refers to the need for sifting out the largest grain for use as seed, and mentions that Celsus advises, where the crop is small, to pick out all the best ears and store them for seed; Varro (1912) similarly advises the selection of the best ears for seed. And Virgil (1947:9) says: "I've noticed seed long chosen and tested with upmost care fall off if each year the largest be not hand-picked by human toil."

Indeed, mass selection probably accompanied the introduction of agriculture. Individual farmers or families would carry out their own selection, often growing a number of different varieties of the same crop to use for different purposes and subsequently selecting seed stock from desirable plants for future crops. The term "landrace" is used to describe any cultivated variety that is the product of centuries or even millenia of farmer selection. Even today, scientific plant breeders have noted the adeptness of modern day native peoples at distinguishing among various local varieties of plants and making use of or manipulating these varieties to serve their purposes (e.g., Hawkes, 1983). Historically, differences in crop usage and in ecological conditions and farming practices between farms led to large numbers of agronomically useful varieties and the maintenance of genetic diversity. Indeed, in the region of origin (or domestication) of each crop species, the genetic diversity was enormous.

Plant improvement was restricted to growing crops for local use for many years. An important development in plant improvement occurred in the seventeenth century with the rise of botanical gardens. These gardens, numbering about 1,600 by the year 1800 (Brockway, 1979), were developed by the colonial powers as a way of identifying plants that might have economic value as plantation crops, transferring plant material from one part of the world to another, and developing taxonomies of the world's plants. The initial taxonomies were based on mere similarities and even included fables about the plants (Foucault, 1970). Only with Linnaeus's classification system (still the basis of modern taxonomy) was order imposed on the bewildering diversity of the plant world. Nevertheless, from their inception, the botanical gardens marked a shift from the cumulative trial and error of peasant agriculture to a more rigorous experimentalism. Henceforth, controlled empirical results became the standard for plant improvement.

Through the botanical gardens agricultural materials, collected on expeditions from throughout the world, would be screened for new and useful crops. The most useful crops, if they could be propagated, were then distributed to foreign colonies for mass production. Indeed, it has been argued that a major reason for the pre-eminence of Britain as a colonial power in the eighteenth and nineteenth centuries was the pre-eminence of her botanical gardens, particularly the Royal Botanic Gardens (Kew Gardens) (Brockway, 1979). At that time, control over plants often meant much wealth and power. During this period, the British controlled much of the world rubber and cocoa production.

Since plants and plant products were of such importance to the colonial powers, an emphasis was placed on plant breeding and taxonomy. In a sense, scientific plant breeding and agricultural research on a large scale began in the botanical gardens. The basic process of plant breeding was not, however, drastically changed by the creation of botanical gardens. Selection procedures tended to remain the same, although much effort sought to determine the nature of inheritance.

The nineteenth century saw two theoretical advances in plant improvement. First, Leibig postulated the existence of three basic elements of plant nutrition: nitrogen, phosphate and potassium. Thereafter, it became possible, at least in principle, to manufacture chemical fertilizers to increase per hectare productivity. Of even greater importance was Darwin's theory of natural selection, which was itself based on the plant breeder's concept of domestic selection (Mulkay, 1979).

By the mid 1800s, domestic selection and the idea of pure lines were well known phenomena, as were a number of theories regarding inheritance. At the time of Darwin, most scientists believed that inheritance involved blending. Genetic material from each parent was believed to blend, like paint, in the offspring. Thus, crossing red flowers with white flowers should give you pink flowers.

In the 1860s, Gregor Mendel's work led to further advances in plant breeding. In pea breeding experiments conducted over 22 years, he evaluated 30,000 pea plants. From his evaluations he concluded that certain pairs of factors (later called genes), derived from the male and female plants, determined the traits or characteristics passed from one generation to the next. Subsequently, Mendel developed two laws of inheritance (i.e., the principle of independent segregation and the principle of independent assortment) that later became the cornerstone of modern genetic science.

Mendel presented the results of his research to the National Science Society in Brunn, Austria, in 1865, but the work was greeted by his contemporaries with indifference. This response was due in part to a lack of understanding about the basic nature of inheritance and in part to the perception that his work held only for certain characteristics of peas. At that time, scientists still believed in blended inheritance. Transmission of characteristics or inheritance via the sperm and egg, was first demonstrated in 1860, and in 1868, Haeckel postulated that the nucleus was responsible for heredity. Almost 20 years passed before chromosomes were singled out as the active factors in heredity. Mendel's work was "rediscovered" in 1900, when three plant breeders – H. deVries in Holland, C. Correns in Germany, and E. Tschermak in Austria – all working independently on breeding experiments, realized the importance of Mendel's forgotten work. Following this rediscovery, the fundamental techniques of today's plant breeding were quickly developed. By the 1920s, inbreeding, hybridization, and backcrossing, were standard procedures. Still, the identity of the genetic material was not yet known (Stern, 1970).

The gradual development of genetics which followed during the period between 1900 and 1920 was furthered by the development of statistics. Statistical analysis of the results of breeding experiments permitted a better understanding of the inheritance of simple Mendelian traits, leading to a focus on populations, as opposed to individuals, and the subsequent birth of population genetics.

The analytical techniques of plant breeding changed somewhat in the 1920s with the development of a powerful breeding tool, the use of single cross and double cross hybrids. Coupled with Mendel's work and with statistical analysis, this approach increased the effectiveness of breeding programs and our understanding of inheritance. While the early 1900s saw both theoretical developments in plant breeding and increasingly sophisticated selection procedures, the actual practice of breeding did not change drastically. Much of the physical work done by plant breeders remained the same.

Major changes in the emphases of plant breeding followed the intensification of agriculture. With the advent of agricultural chemicals (e.g., pesticides, herbicides, fungicides) crop production became less diversified, with fewer crop rotations and with reduction in the range of crop uses. Production became highly concentrated in large monocultures with little or no crop rotation. Since chemicals reduced many pest and disease problems, breeders focused more on increasing yields and less on developing resistance to pests and pathogens.

As the plant breeders became more sophisticated, so did the crops they bred. Primitive cultivars and landraces were gradually replaced by modern cultivars which were more the products of human selection and manipulation than of nature. These domesticated crops required more human care and attention. Garrison Wilkes notes, "their survival is keyed to human beings preparing the ground for them, to human beings guaranteeing them decreased competition with other plants; to human sowing of their seed in the proper season, to human protection during their growth period; and finally human beings collecting their seed. The process of domestication has made these plants our captives" (quoted in Witt, 1985:33).

Traditional Breeding Techniques

Today, crop improvement by traditional breeding schemes usually involves five basic steps: (1) produce and/or discover genetically stable variation for "desired" traits (e.g., yield, disease or pest resistance, stress tolerance, nutritional quality); (2) select from among the variants the individual(s) possessing the desired trait(s); (3) incorporate the desired trait(s) into a suitable agronomic background (i.e., into a form that is appropriate to current farming practices); (4) test the new variety over a wide range of habitats and for a number of seasons; and (5) release the new variety (figure 3.1).

Traditional plant breeding focuses on individuals and populations, and primarily utilizes sexual reproduction both to generate and select for useful variability. The basic emphasis is to increase the frequency of a desired trait in a population and/or combine several desirable traits found in different individuals, varieties, or even closely related species into a single individual.

Four basic breeding schemes are employed to improve plant populations. The first is the development of pure lines; it is usually restricted to annuals (plants that live less than one year and only reproduce once) and to those biennials (plants that live for two years) such as cabbage, onions, carrots, or beets which are treated as annuals in agriculture. In this approach, a plant is repeatedly mated with itself or a close relative (inbreeding) for a number of generations until all offspring are completely uniform in appearance. At the point at which all offspring are

Figure 3.1   Plant improvement through breeding requires large areas devoted to field trials, such as this one for soybean cultivars.

identical in appearance to each other and to the parents (which are also identical) one has developed a pure line; all individuals are homozygous and genetically identical.

The second approach involves producing hybrids. Two different pure lines are crossed with each other. The resultant hybrid offspring are genetically identical heterozygotes. Since first generation hybrids may exhibit hybrid vigor, growing more vigorously and yielding more than either parent, this is often a favored breeding strategy.

Selection in open-pollinated varieties is a third approach. The genetically diverse individuals in a population are allowed to breed among themselves. Offspring with the most extreme value of the desired trait (e.g., yield, percent protein, height, amount of vitamins) are selected and planted as the next generation. By repeating the selection process in each generation, the mean value of the desired trait in the population as a whole is gradually increased.

The final breeding scheme is limited to perennials (plants that live for more than one year and usually reproduce more than once) that can reproduce both vegetatively and sexually. Two clones are crossed, the "best" clone(s) (as determined by the breeder's interests) is selected from among the offspring and planted as the next generation.

The degree of genetic diversity maintained by these breeding schemes varies markedly. Only the third scheme, open-pollination, maintains genetic diversity within a population. The other schemes maintain a great deal of diversity between lines, hybrids, or clones, but within a particular line, hybrid, or clone, all individuals are, in fact, genetically identical.

Both wheat and tomatoes, the subjects of the case studies in the next two chapters, are predominantly inbreeders. In other words, the vast majority of the flowers are self-pollinated. Rates of outcrossing (percentage of all pollinated flowers for which the male and female parents are different plants) vary from 0.5 percent to 4 percent in temperate zone tomatoes (Rick, 1978), and from 1 percent to 5 percent in wheat. Consequently, wheat and tomato breeding fall primarily under the first scheme (pure or inbred lines). However, there is considerable interest, particularly among private breeders, in the use of hybrids.

For inbred lines, the three major variants of breeding practice are pedigree, bulk, and single seed descent. All three practices begin in the same way: a cross is made between two different pure lines. (These crosses invariably are done manually because the plants normally only inbreed.) The first generation offspring (the F_1) are all hybrids which are allowed to self-pollinate and produce considerable variability in the second generation offspring (the F_2). With inbreeding in subsequent generations, this variability is quickly resolved into a number of pure lines. Among these numerous pure lines the breeder hopes to find those that are "transgressive", that is, superior to either of the initial parents for the desired traits.

The three practices differ in the treatment of the F_2 and subsequent generations. In pedigree breeding, considered the classical technique, the plants are widely spaced in the F_2 plots. Single plants are selected and all the seeds from each plant are grown in the same row. In subsequent generations, the best single plants within each row are selected to represent the next generations. As the various lines approach homozygosity, the best rows are selected for further propagation. Their offspring are grown in plots and the best families (i.e., plots) are selected for further propagation.

At the opposite end of the spectrum is single seed descent. With this procedure, rather than selecting a single plant, a single seed is taken from each plant of a large sample. The plants are often grown in crowded conditions in the greenhouse, with two to three generations per year, since the objective is to move through the early generations as quickly as possible with no selective pressure. Once a reasonably high level of homozygosity is reached, usually by the F_4-F_5, a large number of lines, one per plant from the preceeding generation, are established in outdoor plots.

In bulk breeding, which is closer in approach to single seed descent than to pedigree breeding, all plants are harvested, the seeds mixed in bulk, and a random sample used to grow the next generation. Usually there is no selection or only minimal selection in the early generations; any early selection is restricted to characters that are highly heritable and easily handled.

All three breeding schemes conclude, usually at F_5-F_8, with intensive evaluation (quality and yield trials), selection of the emergent lines, and the beginning of purification. By the F_7, only a few genotypes have survived as pure lines. These lines are then tested in a wide geographical area and during a number of seasons to estimate stability over time and space. If the superior lines show relative spatial and temporal stability in yield, their pure stock is multiplied, named, and released, usually between the F_{10} and F_{12}.

Which breeding scheme is preferred? Pedigree breeding, due to the early selection, has the advantage of greatly reducing the number of lines, compared with bulk and single seed descent schemes, that need further evaluation. Although this approach may greatly decrease the amount of work done in later generations, it also decreases genetic variability in the early generations, a potential problem since empirical work indicates that selection in early generations is more efficient for traits with high heritability (i.e., offspring tend to be just like the parents for the given trait) and inefficient for traits with low heritability (Casili and Tigchelaar, 1975). Traits controlled by many genes, such as yield, usually have low heritability, while traits controlled by a few genes have high heritability. Since lines with transgressive yields would occur only rarely, the greater the number of lines brought through early generations, the higher the chances of a transgressive line being present. Thus, when breeding for traits with low heritability, preservation of genetic diversity in early generations must be counterbalanced with reduction of the number of lines that need further evaluation. Consequently, a combination of these schemes is often used.

In addition to these three major breeding schemes, two other breeding techniques, the backcross method and recurrent selection, are also employed. The backcross method is used primarily to introduce one gene or a few genes from closely related species or from a wide intraspecific cross. This method often succeeds in improving one character or a few characters within a variety, especially if the character is simply inherited. Characters controlled by single dominant genes, as with resistance to some diseases and pests, are particularly amenable to this method.

The principle behind the backcross is straightforward. A cross is made between a donor parent (which contains the gene/character to be transferred) and a superior parent (usually a variety with proven agronomic value which lacks, for example, disease resistance). If the F_1 and subsequent generations are always crossed with the superior parent, the donor parent's genetic contribution will be halved in each succeeding generation, until it becomes vanishingly small. If, however, the donor parent contributes an easily selectable character, it may be maintained by selection in spite of the backcrossing, that is, in each generation one saves those progeny which exhibit the desired character and crosses them with the superior parent. The result is a line that is nearly identical to the superior parent, except for one or a few genes from the donor, plus any other genes tightly linked to the transferred gene(s). In practice, a breeder usually backcrosses for four to eight generations and then self-pollinates the plants for several generations to make the selected character(s) homozygous. For both wheat and tomatoes, the backcross method is used primarily for transferring traits such as disease resistance either from different varieties or even different species.

Of the various breeding methodologies, the backcross method most closely approximates the potential outcome of the application of biotechnology: ultimately only a few genes are transferred. This predictability of results leads breeders to its use. Furthermore, the relative precision of the technique means that less extensive testing is required before a new variety is released, fewer plants are needed for the procedure, and breeding can be successfully accomplished under conditions other than those in which the variety will be used.

In contrast, recurrent selection involves alternating cycles of selection and crossing. After crossing two inbred lines, the F_1 is allowed to self-pollinate. The "best" plants are selected from among the F_2 and crossed with each other, which starts the second cycle of selection. East and Jones (1920) first introduced selection within inbred lines and demonstrated its effectiveness in increasing the protein content of corn. This approach is currently employed in corn breeding. Its use in tomato and wheat breeding is limited by the labor input needed to outbreed inbreeders. (To prevent self-pollination and allow subsequent cross-pollination, the male parts of the flower are physically removed before self-pollination occurs. Pollen from another plant can then pollinate the emasculated flower). Hand emasculation of flowers is very labor intensive. Nonetheless, this approach has been successfully applied to wheat breeding. In Minnesota, grain protein percentage in hard red spring wheats was increased by recurrent selection (Loffler et al., 1983).

As mentioned previously, there is a great deal of interest, primarily in the private sector, in developing F_1 hybrids. While hybrid production makes biological sense in cases where hybrid vigor is pronounced and/or where hybrids have been bred for more stable yields over a range of environmental conditions, hybrids are attractive to industry primarily for economic reasons. First, yields of hybrids may exhibit a lower

genotype–environment interaction than do nonhybrids primarily because they have been selected for precisely that trait (R. Lewontin,
personal communication, 1989). Hybrid yields will probably be more
stable over a wide range of environments, and thus a wider geographical area, than yields of nonhybrids. When this is the case, the potential
geographical market is larger for a hybrid than a nonhybrid variety.
Second, profit margins are higher for hybrids than for open-pollinated
varieties. Third, farmers must return to buy hybrid seed each year, since
hybrids do not breed true. Fourth, hybrids are easier to protect from
theft or duplication than pure lines since the parental lines remain
unknown.

History of Molecular Biology and Molecular Genetics

As noted above, conventional plant breeding is an organismic science,
which effects genetic change through conventional sexual means and
attempts to understand what constitutes life at the level of entire plants
and plant populations. In contrast, molecular biology and molecular
genetics are highly reductionistic sciences, which utilize molecular tools
to transform plants genetically and attempt to understand what constitutes life at the level of molecular structures and functions, that is, to
reformulate biological properties and functions in terms of molecular
structure. They have served to promote a shift in what counts as
knowledge, in what features of living matter biology should seek to
examine and explain, and in what is acceptable or interesting as an
explanation. Molecular biology and molecular genetics, from whence
the tools known as biotechnology have emerged, have their origin
outside the plant sciences. Molecular biology began with the study of
simple and often single celled organisms. In fact, for decades, the
favorite organism of molecular biologists was a common intestinal
bacterium, *E. coli*. Moreover, molecular biology is a twentieth century
development based until recently far more on the concerns of apparently sequestered scientists than on the demands of clients. Molecular
biology first approached the everyday world through its impact on
simple disease causing organisms (e.g., bacteria and viruses). Only very
recently have its tools been brought to bear on organisms as complex as
plants.

Unfortunately, the social history of molecular biology and molecular
genetics, as opposed to internalistic accounts of its intellectual development, still remains to be written. However, there are certain features of
molecular genetics that both set it apart from and link it to plant
breeding. There are also certain well-documented episodes that appear

to support our hypothesis that molecular genetics, like plant breeding, may rightly be framed within the "economic" model of science described in the previous chapter. With this caveat in mind, we turn to the history of molecular genetics.

Molecular biology and molecular genetics can be seen as furthering a trend, begun by Descartes, toward reductionism. Descartes introduced the idea that nature can be viewed in a mechanistic way. While he was more concerned with the physical world, his precepts have been widely adopted within the biological sciences as well. Living things can be seen as machines constructed from physico-chemical parts that are fundamentally molecules and atoms. If living things are like machines, then by studying the properties of the *smallest* pieces (atoms and molecules) of the machine and how they fit together, we can understand the whole machine. In the reductionist paradigm, the whole is simply equal to the sum of the parts, and the smallest parts at that.

From this perspective, to learn about life one needs to study smaller and smaller structures. Populations are made up of individuals, which are made up of organs, which are made up of tissues, which are made up of cells, which are ultimately made up of molecules and atoms. Scientists seek to explain the behavior of the higher levels of organization in terms of the lower levels of organization. Such a mechanistic conception of nature reduces the properties of life to the properties of matter in motion, that is, the laws of physics and chemistry. For the geneticist H. J. Muller, life was reduced to the problem of gene structure. Similarly, James Watson, one of the co-discoverers of the structure of DNA, believed that life could be reduced to an understanding of chemical laws. He wrote that

> ... we see not only that the laws of chemistry are sufficient for understanding protein structure, but also that they are consistent with all known hereditary phenomena. Complete certainty now exists among essentially all biochemists that the other characteristics of living organisms ... will all be completely understood in terms of the coordinative interactions of small and large molecules ... further research of the intensity recently given to genetics will eventually provide man with the ability to describe with completeness the essential features that constitute life. (Watson, 1970:67)

J. D. Bernal, whose work on protein structure contributed to the development of a molecular approach, observed that "the appearances of life are biological and scientific consequences of the determination of intimate structure down to molecular and atomic levels" (1967:13). He even defined life as "the partial, continuous, progressive, multiform and conditionally interactive self-realization of the potentialities of atomic electron states" (1967:14). This reductive approach has been very important in the development of molecular biology and molecular genetics and contrasts with traditional plant breeding.

On the other hand, molecular genetics and plant breeding share a view of the role of biology that sets them somewhat apart from their colleagues in other branches of biology. Historically, both have been concerned not merely to understand life but to control it, not merely to understand nature but to create a second nature. Pauly (1987) has recently traced this concern for control in molecular biology to the work of Jacques Loeb in the early twentieth century.

Loeb was an experimental biologist who firmly believed that experiments should be designed for the purpose of controlling nature, rather than merely understanding it. Importantly, one of his mentors was the plant physiologist Julius Sachs, who worked at Berlin Agricultural College. Loeb's biographer, Phillip J. Pauly (1987:40) goes so far as to say that for Loeb, "Successful experimental control was functionally equivalent to scientific explanation." Many of Loeb's contemporaries in biology couldn't understand his apparent obsession with control and his lack of interest in analysis. Moreover, Loeb realized the potential industrial applications of his approach, though he had little personal interest in them. Perhaps no one could sum it up better than Loeb himself: "The idea is now hovering over me that man himself can act as a creator even in living nature, forming it eventually according to his will. Man can at least succeed in a technology of living substance" (quoted in Pauly, 1987:51). Loeb began to work at the Rockefeller Institute in 1909 and helped Abraham Flexner to reorganize it so as to emphasize research that was reductionist and oriented toward control. One of Loeb's students, Hermann Muller, as noted below, used X-rays to induce mutations in *Drosophila*. These events in turn helped set the stage for the rise of molecular biology.

Moreover, the interest in genetics that followed the "rediscovery" of Mendel in 1900 was itself motivated by a desire for biology as control. The plant and animal breeding communities had already put control high on their list of values and welcomed the newly developing field of genetics as an ally. Physicians, whose explanations and therapies were often weak and ineffective, also welcomed it. Additionally, genetics was inextricably intertwined with eugenics through the 1920s. As Rosenberg (1967:36) notes:

Not only within the medical profession, but in the minds of most students of human genetics, social and utopian aspects were foremost. Hence the vogue of eugenics, begun as a movement before 1900, but increasing enormously in influence during the first two decades of this century. Faith in the new knowledge of heredity as a tool of social engineering was not limited to the incompletely educated or the obsessive.

Even Karl Pearson believed that mental retardation could be easily eliminated through the emerging cellular biology and molecular genetics. Similarly, William Bateson thought that racial betterment would soon be possible (Rosenberg, 1967).

The idea of human betterment was also linked to the progressives' ideal of the rationalization of social life (Kohler, 1980). While it would be a gross overstatement to equate progressivism and eugenics, it is fair to say the two movements overlapped and, until the eugenicists moved toward an alliance with fascism, it was not uncommon to find persons who espoused both philosophies. The Rockefeller Foundation's program to support science, developed in the late 1920s under Warren Weaver, reflected each of the currents of the day: progressivism, an acceptance of some of the tenets of eugenics, and reductionism. The Great Depression of the 1930s aided the Foundation program by increasing the urgency of programs for "social betterment", and by putting a damper on progress in the physical sciences which were blamed by many people for the overproduction of the depression. Foundation officials felt confident that the new program in what would become known as molecular biology would serve these diverse, yet complementary ends.

Robert E. Kohler (1980:250) described the changes in Foundation policy from its inception in 1913:

From 1913 to about 1921, policy was to aid college teaching and practical application of science, such as public health programs. From 1922 to 1929, emphasis was put on scientific and professional education, especially medical; in the 1930s and 1940s, a new policy focused on aid to individual research; and from 1950, emphasis was put again on a more practical aim in agricultural science – breeding miracle grains for the "green revolution".

The ideas of the Foundation were not new. Obviously, progressivism was well-established in professional circles in the 1920s, while eugenics was seen as an exciting new field by both the general public and many scientists.

Abir-Am (1987) has traced the idea of a physics-based biology to the informal meetings of a group that became known as the "Biotheoretical Gathering". This group of about a dozen Europeans met through the 1930s and attempted to formulate a new foundation for biology. Weaver was informed of their activities and journeyed to Europe to find out more about them. "Their program, named 'mathematico-physico-chemical morphology', revolved around a joint perspective of the 'exact sciences' on the problem of biological organization, initially at the supra-cellular level of morphogenesis and later at the sub-cellular or molecular level of chromosome and protein structure" (Abir-Am, 1987:10). Weaver eventually provided financial support for their activities as well as for similar groups in the US.

In an effort to explain the new funding policies of the Rockefeller Foundation to the Board and to the general public, Weaver hit on the term "molecular biology", which he used in the 1938 annual report. Between 1932 and 1957 the Rockefeller program directed by Weaver pumped over $90 million into molecular biology (Abir-Am, 1982). A variety of American and European proponents of this "new biology", including Bragg, Morgan, Muller, Schrodinger, Delbruck, Pauling, and Crick, were funded under a number of Rockefeller programs. Though Abir-Am (1982) argues that many of these persons were not funded from the molecular program at Rockefeller, this appears to be of little import, for as Kohler (1980:250) has asserted, the name of the program changed over time "but the idea behind it was constant." Moreover, the largesse of the Foundation had the desired effect of increasing the prestige of the scientists involved (who had ventured beyond the conventional disciplinary boundaries) and of creating training centers where the next generation of biologists could be trained in the new techniques and theories rather than the older natural history approach.

Weaver's program at Rockefeller was important in another way as well. Until its development, scientists usually received institutional funding as in the experiment stations. Weaver set out deliberately to fund only those who appeared to excel in the new biology and to fund them through fairly specific grants. This is not to say that Weaver or the Foundation directed scientists in their day-to-day pursuits, but that the program encouraged those whose goals appeared consonant with those of the Foundation and who responded to what was then a new form of research management. This meant the development of "Big Science" in which large, well-equipped laboratories showed the way for smaller labs that were less able to compete. It also provided the model for the postwar programs of NIH and NSF which emphasized competitive grants and which soon eclipsed the Rockefeller programs in size and scope (Kohler, 1978; Yoxen, 1981).

While some might see the Foundation program as investment in basic research, "Foundation thinking did not encourage a sharp distinction between pure and applied science but between applicable and not yet applicable. All science was ultimately useful, but to be used it had to be perfected as a science, for its own sake" (Kohler, 1980:261). The results of Weaver's program, it was argued, would provide inputs into the more immediately applicable programs of the Foundation in the areas of medicine and public health.

No one could have summed up the Foundation's program any better than did Weaver in a presentation to the Board in 1934:

The challenge of this situation is obvious. Can man gain an intelligent control of his own power? Can we develop so sound and extensive a genetics that we can hope to breed in the future, superior men? Can we obtain enough knowledge of the physiology and psychobiology of sex so that man can bring this pervasive, highly important, and dangerous aspect of life under rational control? Can we unravel the tangled problem of the endocrine glands, and develop, before it is too late, a therapy for the whole hideous range of mental and physical disorders which result from glandular disturbances? Can we solve the mysteries of the various vitamins . . . ? Can we release psychology from its present confusion and ineffectiveness and shape it into a tool which every man can use every day? Can man acquire enough knowledge of his own vital processes so that we can hope to rationalize human behavior? Can we, in short, create a new science of Man? (quoted in Kohler, 1980:263).

Of course, none of this should be construed as a conspiracy or even a direct attempt to insert a particular ideology into biology. Weaver, as a result of his own and the Foundation's conservatism, occasionally failed to provide funds to individuals who were later to make key contributions to the field. Even within the Foundation there were considerable differences of opinion as to whether Weaver's program should be pursued. And scientists often discovered things that Weaver and the Board of the Foundation could never have imagined. In other words, the new science emerged out of negotiations between scientists (as described below) and with Weaver's program at Rockefeller.

Nevertheless, the Rockefeller programs reoriented biology in new directions, toward goals that might well have remained peripheral if the Foundation had not been there to support them. (In this sense the Foundation programs were no different from those established by the state to support plant breeding. It was agreed by those who had the necessary political linkages that plant breeding programs would lead to more rational farm practices; the precise paths taken by breeders remained largely within the province of the breeding community itself.) Finally, even if the early molecular biologists and geneticists ostensibly were uninterested in practical matters, the fundamental concern in the new science with control offered new hopes to important constituencies within the business and professional community, medicine, and the general public.

Like other sciences, the *intellectual* origins of molecular biology and molecular genetics are multiple and diverse. Still, three major origins,

(1) the biochemical and physical, (2) the genetic, and (3) the structural, can be identified (Becker, 1986). While these perspectives have their own historical origins, significant intertwining among them has occurred during the last 50 years, particularly after 1953 with the discovery of the double helix structure of DNA and the proposal of the Central Dogma.

Biochemical and Physical Origins

This approach to molecular biology was pioneered by two physicists, Niels Bohr and Erwin Schrodinger. Bohr, in a speech entitled, "On light and life", delivered in 1932, presented the idea that the workings of living systems may not be understandable in terms of conventional physics, and suggested that a new physics might be necessary to understand life:

> The recognition of the essential importance of fundamentally atomistic features in the functions of living organisms is by no means sufficient for a comprehensive explanation of biological phenomena. The question at issue, therefore, is whether some fundamental traits are still missing in the analysis of natural phenomena, before we can reach an understanding of life on the basis of physical experience (Bohr, 1933:421)

According to Bohr, the study of biology might yield new insights into physics. As a result of this essay, Max Delbruck, one of Bohr's students, left physics to study biology. Inspired by the vision of finding new physical laws in biology, Delbruck concentrated on the fundamental problem of inheritance of variation. A coauthor of the influential paper, "On the nature of gene mutation and gene structure" (Timofeef-Ressovsky et al., 1935), Delbruck asserted that genetics was the biological field in which new physical laws might be found. He went on to note the special characteristics of genes and attempted to relate their physical structure to the production of stable variations (i.e., mutations). Many molecular biologists view this paper as the first seriously to address the connection between gene structure and mutation.

Later, Delbruck, along with Luria, Doermann, and Hershey, became one of the founders of the phage group. From 1937 to 1948 they studied bacteriophages – extremely simple organisms (more primitive than bacteria) – whose high rates of reproduction made them ideal for the study of biological self-replication. Their work focused on genetic information flow as opposed to structure, and ultimately suggested that DNA might be the genetic material.

At that time (1937), geneticists thought that protein was the genetic substance. This view was based on evidence that chromosomes were composed of "nucleoprotein", a complex of nucleic acids (specifically, DNA) and proteins, and that the autocatalytic agents of a virus were nucleoproteins (Bawden and Pirie, 1937). Nucleic acids appeared to have a simple structure compared to the complex structure of proteins. While nucleic acids contained only four different types of bases present in approximately equal quantities, proteins were both chemically diverse, containing some 20 types of amino acids, and often arranged in a complex three-dimensional structure. Consequently, nucleic acids seemed too uniform to account for the diversity of the products produced by genetic material. Indeed, nucleic acids were believed to play a minor role in gene replication: "Nucleic acids are necessary prerequisites for the reproduction of genes.... they are probably necessary for the multiplication of self-producing protein molecules in general" (Caspersson and Schultz, 1938:295).

In 1944, three scientists at the Rockefeller Institute made the important discovery that the hereditary properties of pneumonia bacteria could be specifically altered by the addition of DNA, suggesting that nucleic acid could carry genetic specificity (Avery et al., 1944). This finding modified the nucleoprotein theory of the gene; both protein and nucleic acid might determine genetic specificity. Only after Watson and Crick's 1953 paper on the structure of DNA were nucleic acids seen as the sole genetic material (Watson and Crick, 1953a).

Schrodinger, one of the formulators of quantum mechanics, also became interested in biology as a result of Bohr's essay. In 1945, his influential book, entitled *What is Life?*, popularized Bohr's more esoteric view for a wider audience. This book presented biology as a new frontier which could yield new insights into physics. As Schrodinger states, "from Delbruck's general picture of the hereditary substance it emerges that living matter ... [does] not [elude] the 'laws of physics', which, ... once they have been revealed will form just as integral a part of this science as the former" (quoted in Stent, 1968:393). This interpretion of genetic biology in physicist's terms led to a wave of physicists, including Benzer and Crick (co-discoverer of the structure of DNA), to enter biology. For Schrodinger, the real problem was the physical basis of genetic information. Besides the popularization of Bohr's views, his book offered two major insights which influenced the intellectual development of future molecular biologists, including Crick, Watson, Luria, and Benzer. The first, a proposal about the structural nature of genetic information, suggested that a chromosome was an "aperiodic crystal" which would preserve the structure of the gene in the face of outside influences from the cell. The second insight was that of the "code script" which represented the specificity of the genetic material, its information. This was the first intimation of a coding problem associated with genetic information.

Genetic Origins

A number of the major contributors to the field of molecular biology have written historical accounts of the origin of the field (Watson, 1968; Stent, 1968; Cairns et al., 1966; Pauling, 1970). These works stress the early importance of G. Mendel and F. Miescher in the 1860s, and the chromosome theory of heredity formulated by Walter Sutton in 1902. Other major contributions to the early development of gene theory were made by a group of scientists under the leadership of T. H. Morgan at Columbia University. Known as the Drosophila group, these scientists studied *Drosophila*, a genus of fruit flies. From 1910 to 1915, the Drosophila school further developed the chromosome theory: the idea that physical entities, called chromosomes, carry the hereditary material arranged in a linear fashion. During the 1920s, the same group investigated the mechanism for intergenerational transmission of the particles of inheritance (genes).

Of the scientists in the Drosophila group, H. J. Muller was the most reductionist and the most theoretically inclined, one who "gave full rein to his *a priori* thinking ... and metabolized every new fact into theory" (Carlson, 1971:158). Morgan, on the other hand, was a staunch empiricist with holistic sympathies who disliked theorizing, and was contemptuous of what he called "unbridled speculation". Indeed, before 1910, Morgan's empiricism made him one of the main critics of the chromosome theory. However, as a result of the work of his Drosophila group, he became a major advocate. Morgan and his colleagues were given the major credit for the development of the chromosome theory and Morgan received a Nobel Prize in 1933 for this work.

Muller was an important source of the hypotheses about the chromosome mechanism. Because of his strong beliefs in materialism and reductionism, Muller viewed the gene as a *physical* entity which was capable of mutation and which formed the basis of life. In his book *The Gene: A Critical History*, Carlson (1966) emphasizes the central role of the reductionist concept of the gene in the development of classical genetics and, in a later paper (Carlson, 1974), describes Muller as the main proponent of this gene concept.

These materialist and reductionist views were important because, for the first decade of the 20th century, the prevailing mood in biology was one of essentialism and holism. At the time, holistic, or "vitalist", theories were prominent, in part because of the complexity of living organisms. Furthermore, it was generally assumed that any theory of heredity needed to explain both how hereditary factors were transmitted from generation to generation and how they were transformed into characters. Much speculation existed both about the nature and the intergenerational transmission of the hereditary material. Most theories were forms of biological preformationism, which states that development consists of the enlargement of preformed tissues and organs. An early example of a biological preformist theory was the notion that a miniaturized version of the adult organism, a homunculus, exists in the head of each sperm cell, the egg cell simply containing nutrients to feed the growing embryo. The advent of microscopes discredited this theory. Other preformationist theories continued to flourish, however, such as the concept of pangenes and Darwin's theory of pangenesis. Darwin believed that each tissue or organ of the parent contributed some sort of model, or "pangene", which was incorporated into the sperm or egg, where it guided development in the offspring so as to produce a duplicate of the organ from which it came. The theory of pangenesis was decidedly Lamarckian in character, as changes in the tissues would be transmitted to the pangenes and, through them, to the next generation. In the late 1800s, DeVries removed the Lamarckian weaknesses in Darwin's pangenesis by offering a theory of "intracellular pangenesis".

With the "rediscovery" of Mendel's work, such theories of pangenesis were displaced by debates over Mendel's "factors". William Bateson, one of the most prominent geneticists of the first decade of the 20th century, dominated the period of rediscovery and introduced the terms "alleomorph" (abridged later to "allele", by Shull) and "unit character" for the hereditary units. By 1909, Wilhelm Johannsen, a German botanist, had coined the term "gene", truncated from DeVries's pangenes for unit character (Shull, 1909). Although Morgan and his students still used the term "factor" in their 1915 text *The Mechanism of Mendelian Heredity* (Morgan et al., 1915), by 1917, "gene" was in common usage.

Debate about the "Mendelian factors" in the early twentieth century centered around their structure and their relationship to both mutations and phenotypic characters. Three different approaches to the study of heredity predominated: Mendelian, Lamarckian, and biometrical approaches. The Mendelian approach encompassed numerous competing factions that were united by their belief that the factors obeyed Mendelian transmission ratios. Morgan and Muller maintained that factors were physical particles arranged linearly on chromosomes, determined morphological characters without being isomorphic to them, and stably reproduced their own mutations.

Others including DeVries, Bateson, Castle, Johannsen, East, and Goldschmidt opposed the Morgan–Muller chromosome theory. DeVries, a prominent Dutch plant breeder and a rediscoverer of

Mendel's work, articulated a synthesis of intracellular pangenesis, and his mutation theory, an alternative to Darwinism, proposed the origin of species by singular mutation. Bateson, another major critic of the chromosome theory, also supported biological preformationism. He believed in a direct relationship between the actual physical hereditary material and the phenotype of the organism, specifically in a one-to-one relationship between factor and character such that observable changes in factors were associated with observable changes in characters. "If the chromosomes were directly responsible as chief agents in the production of the physical characteristics, surely we would expect to find some degree of correspondence between the differences distinguishing the types, and the visible differences of number or shape distinguishing the chromosomes" (Bateson, 1909:271).

Given this perspective, chromosomes which were composed of chromatin, a mixture of protein and nucleic acid, could not contain the genetic material. They appeared chemically to be relatively homogeneous, and thus lacked the structural differentiation needed to code for the wide variety of an organism's characters. Thus, in a review of The Mechanism of Mendelian Heredity, Bateson stated: "The supposition that particles of chromatin, indistinguishable from each other and indeed homogeneous under any known test, can by their material nature confer all the properties of life surpasses the range of even the most convinced materialism" (1916:453). Chromosomes were mere epiphenomena, not the hereditary material. While Bateson believed in the physical nature of factors, to think of these factors as beads strung linearly along the chromosome seemed simplistic reductionism. He viewed genetic factors as properties of the cell as a whole, or later the chromosome as a whole, rather than as separable individual particles.

The views of Castle, a British developmental geneticist and staunch opponent of the chromosome theory, were similar to Bateson's. Castle hypothesized a one-to-one relationship of character and factor and, consequently, strongly attacked the idea of the stability of the factors, suggesting instead that genes are altered each generation through either selection or a number of other mechanisms. Wilhelm Johannsen, another major critic of the chromosome theory, was an experimentalist wary of speculative theory. Like Bateson, Johannsen believed that genetic factors were a property of cells as a whole. Unlike Bateson, however, Johannsen viewed genes as mental constructs, not necessarily as physical particles. To Johannsen, "the genes were still primarily accounting units [Rechnungseinheiten], an 'expression for realities of an unknown nature, but with known effects'" (Roll-Hansen, 1978:206). The geneticist E. M. East held views similar to those of Johannsen, conceiving of genes in an idealist fashion, solely as mutational symbols with no concrete reality (East, 1912).

The views of those Mendelian factions opposing the Morgan-Muller chromosome theory had a predominantly antimaterialist and/or anti-reductionist character. William Coleman (1970) argues that Bateson rejected the chromosome theory in part because of his morphological training and his general antimaterialist metaphysics. Bateson's approach was one of physical holism, in which a whole has properties that cannot be explained from its parts. Thus, he urged biologists to study the whole organism as the fundamental entity. While he sought to solve problems of heredity through the direct application of physical principles, he employed the antimechanist physics of the late nineteenth century, which dealt in terms of fields rather than particles (i.e., atoms). Similarly Johannsen's beliefs were based on physical holism. Johannsen, a staunch experimentalist, was greatly influenced by the work of the French physiologist Claude Bernard who advocated an antireductionist methodology (Roll-Hansen, 1978). Finally, Goldschmidt's theory of gene action was explicitly holistic as well as physiological (Goldschmidt, 1916). His opposition to the Morgan-Muller gene theory was due both to his strong philosophical materialism and to his belief that the study of inheritance could not be divorced from physiological and developmental phenomena. In his view, any concept of heredity should explain both gene transmission and gene action (i.e., how the gene is translated into a phenotype).

The other two approaches to the study of heredity, the Lamarckian and the biometrical, were definitely non-Mendelian. While the Lamarckian approach stressed the inheritance of acquired characters, the biometrical approach, championed by Pearson, a co-founder of the field of biometry, attempted to deepen Darwin's idea of gradual speciation through the natural selection of minute variations. A strong believer in phenomenalism and instrumentalism, Pearson's total rejection of Mendelian concepts has been linked to his philosophical and ontological position (Norton, 1975). Thus, "Pearson held that all meaningful statements could be analyzed into assertions about actual and potential sense impressions" (Norton, 1975:88). Consequently, he thought the task of science was to describe new laws which, in keeping with positivist theories of the time, would use more parsimonious language to describe the facts.

Since Mendelism failed to generate a few simple general principles which could explain all the results from breeding experiments, it violated Pearson's phenomenalism and instrumentalism and seemed hopelessly *ad hoc* to him. Furthermore, Pearson had an ontological viewpoint which sought to outlaw homogeneous classes from all branches of science. As a developer of statistics he understood the power of correlation mathematics. He believed that phenomena in nature were rarely linked functionally, pointing to the statistical basis for much of chemistry and physics, and noting that "the conclusions of chemists and physicists are based on *average* experience" and that "no two ... [conclusions] actually agree" (Pearson, quoted by Norton 1975:91). Thus, he argued that science should view nature in terms of correlation and *not* causation. This required emphasis on variation rather than sameness. Any theory that used homogeneous classes as explanations, as did Mendel's, could not be correct.

The publication of *The Mechanism of Mendelian Heredity* by the Drosophila group (Morgan et al., 1915) represented both the foundation of classical genetics, as well as the death knell for the opposing schools of thought. It provided firm support for both the chromosome theory and the reductionist view of the gene. In England, Bateson continued to vigorously attack the Drosophila group's theories until 1921, when he admitted, after a visit to Morgan's laboratory, that they might be right after all. In the US, Castle persisted until 1919. DeVries's mutation theory, dubbed an alternative to Darwinism, was discredited by 1919 and neo-Darwinism prevailed (Carlson, 1974). Thus, by the early 1920s, the reductionist view of the gene became the predominant paradigm in genetics.

Pearson's biometrical school, which rejected both the chromosome theory and Mendelism, remained outside classical genetics until the 1930s when S. Wright, R. A. Fisher, and J. B. S. Haldane combined the reductionist view of the gene with a biometrical analysis of Darwinian variation, thereby founding the field of population genetics. This combination of reductionist gene theory, biometrical statistics, and Darwin's theory of evolution by natural selection has been called the synthetic theory of evolution or the neo-Darwinian theory and was considered a major advance in evolutionary theory.

Structural Origins

Scientists in the structural school held that a cell's physiological functions could be understood only in terms of the three-dimensional configuration of its components. Thus, they placed emphasis on determining the three-dimensional structure of molecules. X-ray crystallography, a primary technique of the structural school, involved exposure of the substance to X-rays followed by analysis of the resultant X-ray diffraction pattern to obtain precise information on the three-dimensional arrangement of atoms in crystal molecules. Originally invented by a British father-and-son team, W. H. and W. L. Bragg, and used in 1912 to determine the structure of sodium chloride (Bragg, 1913), this approach led to the founding of a British school of crystallography. Increasingly complex molecules were studied with this new technique, including the structure of complex inorganic silicate molecules by the Nobel laureate Linus Pauling (1962). By the late 1930s the technique was applied to biological molecules by two of the Braggs's students, J. D. Bernal and W. T. Astbury. Bernal's laboratory began structural analysis of proteins in the mid 1930s, while Astbury worked with nucleic acids. Structural studies in Bernal's laboratory of hemoglobin, a blood protein, and myoglobin, a muscular protein, proceeded slowly. Although Max Perutz began his hemoglobin research in 1937, and Andrew Kendrew initiated his studies of myoglobin, a protein four times smaller than hemoglobin, in 1947, it was not until the late 1950s that any major results were obtained (Kendrew et al., 1958; Perutz et al., 1960). Inferring the structure of these biological molecules from X-ray diffraction patterns proved difficult since no logical method existed, at the time, for their interpretation.

A major breakthrough occurred in 1951 when Linus Pauling, using diffraction data and a model-building approach, proposed that polypeptide chains (or proteins) were frequently arranged in a helical fashion, specifically an alpha-helix (Pauling et al., 1951). Unfortunately, this proposal did not immediately suggest any new information about proteins, such as how they function or how they are synthesized. However, Pauling was on the right track. In early 1953 he proposed that DNA was composed of a triple alpha-helix (Pauling and Corey, 1953). Soon after, Watson and Crick (1953a) surmised the currently accepted structure of DNA, a double alpha-helix, in large part because of the X-ray diffraction data of Rosalind Franklin (Franklin and Gosling, 1953).

The biochemical studies of Erwin Chargaff (1951) on the base composition of DNA can also be thought of as part of the structural school. Stimulated by the demonstration that DNA might be the hereditary material (Avery et al., 1944), Chargaff wanted to know how the four bases of DNA were arranged. The prevailing hypothesis, the tetranucleotide hypothesis, maintained that the four nucleotide bases were present in equal proportions (i.e., 25 percent) in all DNA molecules. When evidence, including Chargaff's, subsequently demonstrated that the four nucleotides were not present in equal proportions, the statistical polytetranucleotide theory, stating that the proportions of the four bases were random, emerged. Between 1945 and 1950, Chargaff established that base ratios might differ for different organisms, although they did not vary randomly within a species. A set of well-characterized but poorly understood relationships, known as Chargaff's rules, were shown to exist.

The Modern Era: Molecular Genetics

The currently accepted structure of DNA was first proposed in 1953 by James Watson and Francis Crick in a paper entitled, "The molecular structure of nucleic acids: A structure for deoxyribonucleic acid"

(Watson and Crick, 1953a). Combining results from genetic, structural, and biochemical approaches, they guessed the structure of DNA. Originally sent to Europe to learn nucleic acid chemistry, Watson, a student of Delbruck, was intrigued by Maurice Wilkins's (Wilkins et al., 1953) X-ray pictures of DNA and decided to study structural crystallography at Cambridge. Collaborating with Crick, he attempted to introduce genetic reasoning into the structural determination of DNA. The desired structure needed to explain both how DNA could replicate itself, the autocatalytic functions, and how it could induce, or preside over, the synthesis of virus-specific proteins. Their double helix model for DNA structure suggested how replication and even mutation could occur (Watson and Crick, 1953b). This hypothesis subsequently served as the basis for a general theory, the central dogma, whereby DNA serves as a template for its own replication and codes for mRNA which, in turn, codes for protein (Crick, 1958). The central dogma views DNA as the master molecule, with genetic information flowing unidirectionally, from DNA to RNA to protein. Furthermore, genetic material (DNA or mRNA) is viewed (metaphorically) as information; DNA contains a genetic "blueprint" or "program" that is faithfully "transcribed" into a "messenger" molecule (mRNA) which is "translated" into a protein. Thus, even the vocabulary of molecular biology places it within the discourse of the larger theoretical paradigm of information theory (Yoxen, 1983).

This model catapulted DNA into prominence, launching the era of molecular genetics that has since revolutionized all of biology. Within a few years, experimental data supported the hypothesis that genetic information is conveyed by the sequence of DNA base pairs (1956) and that DNA replication involves separation of the complementary strands of the double helix (Meselson and Stahl, 1958). This was quickly followed by the isolation of DNA polymerase I, the first enzyme that synthesized DNA in a test tube (1958) and the discovery of RNA polymerase, an enzyme that polymerizes RNA chains using a DNA template (Kornberg, 1960). Other significant developments in the 1960s included: the discovery of messenger RNA and the demonstration that it carries the information needed to order amino acids in proteins (Hall and Spiegelman, 1961); the use of a synthetic messenger RNA molecule (poly-U) to determine the first letters of the genetic code (Nirenberg and Matthaei, 1961); the realization that genes conveying antibiotic resistance in bacteria are often carried on small supernummerary chromosomes called plasmids; and the isolation of the enzyme DNA ligase, which can join DNA chains together (Weiss and Richardson, 1967).

Early in 1970, H. O. Smith isolated the first restriction enzyme, an enzyme that cuts DNA molecules at specific sites (Smith and Wilcox, 1970), a discovery for which he received the Nobel Prize in Medicine in

1978. Shortly afterward, in 1972, H. Boyer and S. Cohen of Stanford University developed the first practical methods for systematically cloning specific DNA fragments. (This involved the use of DNA ligase to link DNA fragments created by restriction enzymes.) The next year foreign DNA fragments were inserted into plasmid DNA to create chimeric plasmids, which were subsequently functionally reinserted into the bacterium *E. coli* (Cohen et al., 1973). The potential now existed for the cloning in bacteria of any gene. The new recombinant DNA biotechnology was now a reality (Watson et al., 1983).

The decade of the 1970s brought further developments in molecular genetics, including the creation of the first recombinant DNA molecules containing mammalian DNA (Rougeon et al., 1975) and the discovery of split genes (Breathnach et al., 1977); the development of procedures for the rapid sequencing of long sections of DNA molecules (Sanger and Coulsen, 1975; Maxam and Gilbert, 1977); production of the first human hormone, somatostatin, through recombinant DNA techniques (Shine et al., 1977); prenatal diagnosis of the genetic disease sickle cell anemia directly at the gene level (Chang and Kan, 1981); and incorporation of a foreign gene into the cells of a tobacco plant with subsequent transmission of the gene to the plant's progeny in a simple Mendelian manner through the pollen and the egg cells (DeGreve et al., 1982). The next section describes in greater detail these new biotechnologies, particularly those related to genetic transfer in plants.

The New Biotechnologies

In contrast to the techniques of traditional plant breeding, the new biotechnologies utilize molecular genetics and focus on the cellular and subcellular levels. These technologies were originally used to examine the genetics of single-celled organisms. "However, the turning point was in the realization that these discoveries ... conferred upon higher plants many of the attributes that had made microbes so amenable to genetic study" (Chaleff, 1983:676–7). Instead of studying millions of plants in a field one could study millions of plant cells in a petri dish. In the initial excitement generated by this metaphor (i.e., cultured plant cells are microbes), the very metaphoric character of the relationship was lost to some. Chaleff explains: "With recognition of the similarities between cultured plant cells and microorganisms came the expectation that all the extraordinary feats of genetic experimentation accomplished with microbes would soon be realized with plants. But because of the many ways in which cultured plant cells are unlike microbes, these expectations thus far have not been fulfilled" (1983:679). These tech-

Figure 3.2 The new biotechnologies bring plant research indoors.
(Photo courtesy of Jim Collins, Pennsylvania State University)

nologies, which can be divided into two broad classes, culture technologies and genetic transfer technologies, are reviewed below.

Culture Technologies

The culture technologies function at the cellular level and above, and involve regenerating entire plants from protoplasts, single cells, tissues, organs, or embryos. The genetic transfer technologies function at the subcellular level and involve the transfer of one or a few genes between cells, usually of different species. Although the genetic transfer technologies, due to their greater precision, ultimately hold greater promise for the production and manipulation of useful variability in crop plants, the culture technologies represent a limiting step in crop improvement. Following transfer of genetic material into cells, these transformed cells must be regenerated into whole plants before they can be utilized in crop improvement (figure 3.2).

Culture techniques generally involve three basic steps: (1) plant parts (organ, tissue, cell, etc.) are separated into fragments, cell suspensions, and/or protoplasts; (2) cell division and growth into a mass of undifferentiated cells (callus tissue) is induced by placement in an appropriate nutrient and hormone solution; (3) callus tissues are exposed to appropriate conditions for redifferentiation, producing roots and a shoot that grow into a normal plant.

Depending upon the system, a number of variations are possible. In some systems, the callus stage can be omitted entirely; redifferentiation occurs directly from the plant piece. In some species certain cell types can be induced to undergo somatic embryogenesis, where embryos form directly from callus tissue. In general, somatic embryos are more likely to develop into mature plants than unorganized callus tissues. In addition, either haploid or diploid cells may be employed. The techniques involving haploid cells often include a step, such as exposure to colchicine, that induces chromosome doubling. There is much interest in these doubled haploids since complete homozygosity can be achieved in one generation, thereby reducing variety development time. In general, however, the ability to use these culture techniques varies widely with cell types and tissues as well as genotypes.

The potential benefits and uses of these culture techniques are multiple. Since an enormous number of different plants may be grouped in a small area, one can screen a large number of cells for new and useful mutations in a very short time. Additionally these techniques allow access to a greater range of mutations, or variability, since mutants occurring in either somatic (body) cells or sex cells can now be propagated. The natural variability of somatic cells, called somaclonal variation, can be evaluated through culture techniques. In fact, some believe that many new varieties will be developed by the procedure (Evans et al., 1984; Orr, 1985). In tomato, for example, somatic mutations have given rise to genetically altered plants whose offspring stably exhibit the mutations (Evans and Sharp, 1983). Most breeding programs are based on producing variability and then selecting from among the variants. Consequently, in addition to natural somaclonal variations, and those mutations resulting from the culture technique themselves, scientists have induced variability via the use of X-rays or certain chemicals as mutagenic agents (OTA, 1981, 1984). Selection of desired variants can be facilitated by addition of a specific harmful substance, for example a virus, a herbicide, salt water, heat, etc., to the growing media since only those cells resistant to that substance will grow. This approach can greatly reduce the time required to develop disease- or stress-resistant varieties.

These culture techniques are not without their drawbacks. Induced phenotypic mutations frequently are unstable across generations, differ from spontaneously occurring mutations, and rarely have a positive effect on traits linked to productivity. In the tomato example mentioned above, the recovered somatic mutations were previously known single-gene mutations which affected various traits (such as fruit color, male fertility, chlorophyll formation, and growth habit), none of which were associated with higher yield. Much like induced mutations, somaclonal variation represents a shotgun approach to breeding, giving rise to mutations which are nonspecific in nature. The optimistic predictions about the utility of somaclonal variation are reminiscent of the initial predictions made about mutation breeding in the 1950s and early 1960s. Some thought mutation breeding would play a prominent, if not dominant, role in plant breeding. Although these predictions were wrong, in some cases the use of mutagenic agents has proved to be a useful adjunct to more traditional methods. Somaclonal variation may play a similar role in plant breeding.

In general, culture techniques are not as successful with monocots (which include all the basic cereals) as with dicots. The major problem appears to be regeneration of cell suspensions or protoplasts into plants (OTA, 1984). A wide variety of dicots are amenable to cell and tissue culture, with crop plants in the *Solanaceae* (tobacco, tomato, and potato), *Umbelliferae* (carrots), and *Brassicaceae* (mustards) genera being easier to culture (Evans et al., 1984). Why this is so remains a mystery.

A final problem with the culture technologies is the lack of information about the control of gene expression during development, that is, understanding how cells regulate the expression of specific genes at different growth stages. Thus, even when culture technologies are successful, one seldom knows why. Indeed, as reported by the National Academy of Sciences (NRC, 1984:26): "Although an increasing number of plants are yielding to protoplast and other culture techniques, these advances have stemmed as much from guesswork as from science."

Genetic Transfer Technologies

In contrast to culture technologies, genetic transfer technologies function at the subcellular level and can involve the transfer of genetic material between species that are incapable of breeding with each other. The genetic material transferred can be either general or specific. At the general level, somatic cell hybridization (or protoplast fusion) permits the formation of hybrids between widely different plant species. Protoplasts isolated from different plant species and mixed in a culture medium are induced to fuse, either by use of chemicals or an electric field (Cocking, 1983). If the new cell successfully integrates the two nuclei and undergoes cell division, hybrid cells will be formed. If the nucleus of one of the protoplasts is destroyed prior to fusion (usually by radiation), the fusion product, called a cybrid, is a cell with a hybrid cytoplasm (i.e., it contains cytoplasm from two different cells) but a nonhybrid nucleus.

In addition to entire cytoplasms, one can also transfer organelles, such as chloroplasts or mitochondria, by a variety of methods. Liposome transfer, a common method, involves surrounding the organelle to be transferred with a lipid membrane to form a liposome, which can then be fused with a protoplast. If the organelle being transferred is a chloroplast, the fusion product is called a chlybrid; if the organelle is a mitochondrian, the fusion product is a mibrid. Cybrids, chlybrids, and mibrids may become very useful in plant breeding since a number of important traits, including cytoplasmic male sterility, herbicide resistance, photosynthetic efficiency, and mitochondrial efficiency are cytoplasmic and/or organellar factors (Cocking, 1983).

In general, however, hybrids from widely different plants are sought whether to acquire more effective resistance or tolerance to diseases, pests, or stress conditions, or to improve product quality and other characteristics enhancing crop productivity. Some have even proposed the development of hybrids between legumes, which have the ability to fix nitrogen, and nonlegumes (Barton and Brill, 1983), as well as between species with differing photosynthetic efficiencies (Cocking, 1983). There are many difficulties with these methods and, at this time, they can only be applied to a few plant species. When fusing whole protoplasts, the main problems involve obtaining cell division and preventing genetic material from being lost or ejected from the cells during cell division. In general, the more closely related two plant species are, or the smaller the amount of material being transferred, the higher the probability of success. For example, protoplast fusion has been successfully used to develop tomato–potato hybrids (Shepard et al., 1983).

The recombinant DNA (rDNA) technologies are the most precise of the genetic transfer technologies, as only one or a few genes are transferred at a time. The first steps involve isolating and cloning a DNA segment which contains the desired gene(s). Commonly, one exposes the donor DNA to an enzyme (called a restriction endonuclease) which cuts or restricts the DNA at specific base pair sequences, thus generating double stranded DNA fragments. At each end of a fragment one strand of DNA is longer than the other, resulting in unpaired bases. Since the unpaired bases will readily pair with complementary bases, these ends are called "sticky ends". To transfer the genes into other cells requires a vector, often a plasmid. A plasmid is a small circular piece of DNA, separate from the main bacterial chromosome, which is capable of independent replication. Under appropriate chemical conditions, a plasmid may enter a host cell and reproduce itself. If the plasmid is exposed to the same restriction endonuclease as the donor DNA, the bases in its sticky ends will be complementary to those in the donor DNA. Similarly restricted donor DNA and plasmid DNAs are then mixed and treated with an enzyme which rejoins broken DNA molecules. This procedure generates many repaired plasmid molecules, some of which will also contain inserts of the donor DNA. These recombinant plasmids are then mixed with bacterial cells under chemical conditions which facilitate the uptake of the plasmids by the bacterial cells. When the recombinant plasmids replicate, the inserted donor genes will also replicate. Subsequently, the machinery of the bacterial cells is used to transcribe and translate the donor genes into their respective gene products. In sum, the bacterial cells serve as miniature factories which produce the desired gene products. One can then isolate those DNA fragments that contain the gene(s) of interest. Thus, these rDNA techniques permit isolation of small pieces of DNA and analysis of their gene product. Consequently, these techniques have already revolutionized molecular biology and genetics and have rapidly expanded our knowledge of the genetic structure and function of procaryotes.

The potential for rDNA far exceeds that of protoplast fusion and organelle transfer due to the finer control it provides over which genes are transferred. Traditional breeders transfer whole chromosomes carrying clusters of characters, some beneficial and some deleterious. The real work of traditional plant breeding is to eliminate the undesirable qualities while keeping the desirable ones. In principle, with rDNA techniques one can transfer only desirable genes and thereby decrease the need to sort out the undesirable traits. In addition, the genetic transfer technologies allow transfer of DNA between any two organisms, while the traditional techniques are restricted to sexually compatible species.

In general, genetic transfer technologies are farther from application than the culture technologies. The rDNA techniques presuppose knowledge of the identity and location of the particular gene(s) to be transferred. While over 800 genes have been mapped on the one chromosome of the bacterium, E. coli, the locations of genes in commercially valuable crops are only partially mapped even for the most studied species. For example, in tomatoes, not that long ago only 970 genes had been mapped (Tigchelaar, 1986). Furthermore, since most traits linked to productivity are multigenic in character, the utility of rDNA for introduction of desired genes may be more limited than proposed. Still, these molecular and biochemical techniques may greatly facilitate the construction of chromosome maps for all crop plants (Flavell, 1985). Additionally, transcription and translation are much more complex in eucaryotes than in procaryotes. The majority of eucaryotic DNA does not code for protein; while some of this DNA is involved in the regulation of gene expression, the function of the remainder is unknown.

Another major constraint on the use of genetic transfer technologies is limited knowledge concerning the regulation of gene expression (i.e., understanding how genes are turned on and off at the appropriate times during development). At any given time, only a small percentage of a cell's DNA is being actively transcribed and translated. In addition, there is differential gene expression between cells, with specific genes selectively expressed in certain cells. For example, in tobacco, the petals and leaves each express about 7,000 specific genes, while the ovary and anthers each express about 10,000 specific genes (NRC, 1984). Scientists must be able to control when and where genes are expressed for these technologies to be useful. A gene for herbicide resistance which is expressed in the roots, while the herbicide is applied to the leaves, or which is expressed just prior to flowering or fruit set, whereas the herbicide is needed in the early stages of crop growth, would be of no use to a farmer. Furthermore, in certain cases where regulation of gene expression is understood, regulatory DNA sequences are physically distant from the gene whose expression they control (Schaffner et al., 1985). Thus, in addition to knowing *how* gene expression is regulated, knowledge of the precise location of regulatory sequences is required.

Finally, since plasmids replicate only in procaryotes, vectors must be found to carry foreign DNA *into* the plant genome. Only a few bacterial plasmids are effective in plants. Of these, a plasmid from the soil bacterium *Agrobacterium tumefaciens*, which causes crown gall disease in many plant species, is of major interest (Chilton et al., 1977). This disease is characterized by tumor growth resulting from insertion of genes from a tumor inducing (Ti) plasmid into the plant cell's chromosomes. The gene products of this foreign DNA cause tumor formation and growth. Once its tumor inducing properties are suppressed, the Ti plasmid may serve as a good vector. It has been used to transfer subsequently expressed genes from sunflowers to beans (Netzer, 1983a). Some work has also been done with a plasmid from a related species of soil bacteria, *Agrobacterium rhizogenes*, which causes hairy root disease (Quattrocchio et al., 1986). This plasmid, once its DNA has been inserted into the host plant cell's DNA, induces the growth of roots from the infected cells. However, the use of these two plasmids is fraught with problems, including control of the amount of DNA transferred into the plant cell, control of the site of DNA insertion, transformation efficiency (only a small percentage of cells exposed to the plasmid become infected), and selection of transformed cells, i.e., detecting those cells that have taken up the plasmid and express its gene products and separating them from either untransformed or nonexpressing cells. Furthermore, the plasmids from *A. tumefaciens* and *A. rhizogenes* do not infect monocots.

For monocots and other plants not susceptible to Ti plasmid infection, alternative vector systems or systems for the delivery of free DNA are needed (Gasser and Fraley, 1989). Plant viruses may be ideal candidates for vectors, given their ability to infect plants by injecting their DNA into plant cells. However, a major problem with the viruses studied is the amount of DNA transferred; so far, only pieces of DNA smaller than a single gene have been transferred.

Much interest exists in using transposable genetic elements, also known as the "jumping genes", first discovered in corn by Barbara

McClintock in the early 1960s, as a vector system for cereals (Gerlach et al., 1985). Specific transposable genetic elements, known as transposons, are segments of DNA which can replicate themselves to produce additional copies for subsequent insertion elsewhere in the genome. Since transposons do not move between cells, but only between and within chromosomes in the same cell nucleus, genetically engineered transposons could be used to insert foreign DNA into a plant cell's chromosomes. However, another method such as microinjection is required to introduce engineered transposons into the host cells. In this procedure DNA to be transferred is literally injected into the recipient cell utilizing a microscopic syringe. To reduce problems of plant regeneration of transformed cells, most work focuses on using egg cells, pollen, and young embryos (called fertilized ovules). Successful transformation of cotton by microinjection of DNA into newly fertilized ovules has been reported (Zhou et al., 1983). In addition, the International Plant Research Institute is patenting a vector system for transforming cereals that utilizes a transposon and microinjection technique (Netzer, 1983b).

Recently, systems for the delivery of free DNA have received much attention. While the first of these systems utilized plant protoplasts and relied on physical means for facilitation of DNA uptake, for example, calcium chloride precipitation, polyethylene glycol treatment, or electroporation (Gill and Warren, 1988), more recent work has emphasized introduction of DNA into intact cells or tissues. Development of the "particle gun" or high-velocity microprojectile technology (Klein et al., 1988) has facilitated these efforts.

In sum, for the rDNA techniques to be useful in plant breeding, three basic criteria must be met. First, the specific gene(s) responsible for a given trait must be identified and isolated. Second, the gene(s) must be introduced or transferred into a regeneratable plant cell. Finally, the developmental expression of the gene(s) must be regulated. While this final criterion may be the most difficult to achieve, since it requires an understanding of the mechanisms controlling gene expression, the other criteria also present difficulties.

Traits controlled by multiple genes represent one type of difficulty, particularly when the majority of these genes have not been identified. This is the case for multigene, or horizontal, resistance to insects and pathogens. While horizontal resistance is far more stable than vertical or single-gene resistance, due to the large number of mutations which must occur in the same insect pathogen before the resistance can be overcome, application of the new biotechnologies cannot proceed until all of the genes involved are identified. Consequently, this approach has little to offer over conventional breeding techniques. Much of the plant biotechnological research currently underway focuses on such areas as developing herbicide resistance which is often a single-gene trait (Gasser and Fraley, 1989), or introducing single-gene, or vertical, resistance to insects or pathogens. The latter has little long-run value as both insects and pathogens are capable of mutations that quickly overcome the new source of resistance (Borlaug, 1983). In nature, most effective resistance is multigenic in character.

Another difficulty involves the control of gene expression. While the new biotechnologies may initially effect gene transfer and expression at the subcellular or cellular level, desired characters (e.g., resistance to a pathogen) must ultimately be expressed at the appropriate time in the appropriate plant tissue(s), a more difficult level of regulation to achieve. Furthermore characters expressed at the whole-plant level (e.g., yield; resistance to lodging) may not be amenable to analysis at the cellular level. Until correlates of these whole-plant characters can be found at the cellular level, the new biotechnologies cannot systematically address these aspects of plant improvement.

While the new biotechnologies differ from plant breeding in the two initial steps involved in crop improvement, conventional breeding techniques are needed to establish a desired trait in a suitable agronomic background, and to test it over time and space, before final release. For maximum adaptation, final stages of selection should occur in an environment identical or similar to production conditions. These final steps are the most time consuming for plant breeders: breeders estimate that the final three steps require approximately 75 percent of their plant breeding effort (personal interviews).

Nevertheless, the new biotechnologies may facilitate two major interrelated changes in breeding programs: First is the truncation of both time and space. Initial screening of materials in the laboratory both reduces the space needed to grow thousands of plants and the time needed for those plants to mature. Second is the introduction of greater precision into the breeding process. By bypassing the process of sexual reproduction, it may be possible to insert only desirable genetic material into a particular crop.

Scientists' Origins and Disciplinary Prestige

Today, conventional plant breeders and molecular geneticists, like most biological scientists, make use of a wide array of scientific instruments, publish in and read the genetics literature, use a complex technical language to describe their objects of study, and rely heavily on statistical procedures. However, they differ considerably in terms of both the historical origins of their fields and the methodologies and techniques they currently employ. In addition, they differ markedly in terms of family origin and education and in terms of the reputation of their disciplines in the scientific community.

Family Background

It has been posited that scientists are the basic determinants of the rate of development of science, technology and social institutions. Scientists provide the expertise and knowledge for scientific development. They bring a variety of personal qualities, background characteristics and social histories to the research setting. A growing body of literature and empirical studies has demonstrated the remarkable influence of these person-bound factors throughout the production, interpretation, and evaluation of scientific results (Feyerabend, 1975; Latour and Woolgar, 1979; Busch and Lacy, 1983).

In examining the family origins of scientists and, in particular, agricultural scientists, it is interesting to note that scientists are far more likely to come from farm backgrounds than would be expected by chance. In a survey of 10,000 individuals receiving doctorates from US universities between 1935 and 1960, it was revealed that scientists in all fields were more likely than the general population to have fathers who were farmers or farm managers. However, 49 percent of the agricultural scientists reported that their fathers were either farmers or farm managers, in contrast to only 15 percent of individuals in all fields of science and engineering taken together. This pattern was repeated in a national study of agricultural scientists conducted in the early 1980s. Busch and Lacy (1983) found that 54 percent of agronomists and plant breeders reported coming from farm backgrounds, as opposed to only 26 percent of the agricultural basic scientists, including agricultural molecular geneticists. Finally, of the scientists we interviewed, 29 percent of those engaged in molecular genetics (6 of 21), but 60 percent of the plant breeders (25 of 42), came from farm backgrounds. Thus, plant breeders appear more than twice as likely to come from farm backgrounds than agricultural molecular geneticists. It is unclear what the precise consequences of family origins are for research. However, the national study of agricultural scientists by Busch and Lacy (1983) found that those scientists with farm backgrounds were more likely to address research questions that met client needs and demands, and responded to feedback from extension staff. Those with other family backgrounds were more likely to be committed to scientific ideals. This suggests that those with farm backgrounds are more likely to have closer connections to and identification with farming and the problems associated with it than other scientists.

Disciplinary Goals and Status

The two fields also differ in terms of the mission and status of the disciplines. Plant breeding is considered an applied science with a strong mission – the production of improved crop varieties – while molecular genetics is characterized as a basic science. Although some have noted that the distinction between basic and applied research may be blurred at times (Mulkay, 1977), others (e.g., Levins, 1973) contend that a strong dichotomy between the two exists in the biological and agricultural sciences. It is often argued that *basic* research aims to increase knowledge or understanding regardless of practical benefit, while *applied* research aims to increase knowledge that has specific commercial objectives or applications. Furthermore, applied research is often viewed as the applications of basic science to concrete problems in the everyday world. As a corollary, basic research is seen as having fewer constraints than applied research, with problems (or research topics) chosen based on the investigator's interest and/or on relevance to fundamental knowledge. The direction of research is dictated solely by the investigators and their previous results. For applied research, problem choice is determined by economic interests or a desire to solve a given practical problem, with research direction being constrained by the need to progress toward the solution of a practical problem.

As a result of these apparently different goals, basic science is often seen as more intellectually rigorous as well as more worthwhile and glamorous than applied science. Furthermore, it is assumed that breakthroughs will come more quickly with basic than with applied research. The criticisms of agriculture voiced in both the Pound (National Research Council, 1972) and Winrock (Rockefeller Foundation, 1982) reports echo this perception. Both reports state that agricultural research in the US is in a declining or stagnant state, in part, because the research is too applied in focus. An increased emphasis on basic research, or research at the "frontiers" of science, the reports maintain, is needed to revitalize and enhance agricultural research. In addition, more applied research should be performed by the private sector. Indeed, the Winrock report went on to discuss and suggest structural changes in the land grant university system which would facilitate greater influence of the private sector on land grant university research, as well as shift the emphasis away from applied, and toward basic, research. In part as a response to the criticisms contained in these two reports, and especially the latter, land grant universities have increasingly focused more on high technology and basic research and less on applied research. This change in focus has occurred as part of a larger shift in the purpose and function of land grant universities (Schuh, 1986) over the course of several decades.

To better understand the current changes in the land grant universities, a brief review of their 125-year history is helpful. The land grant university system was established in 1862 to counter the elitism of private universities. The term "land grant" has its origins in grants of government land in the western United States that were given to each state as a form of endowment for constructing a university to provide education in "agriculture and the other mechanic arts". The land grant universities had, as a mission, to serve the needs of rural people. Their purpose was to educate the rural masses and to improve the quality of rural life by doing research on the pressing agricultural and social problems of the day. Scientists were rewarded for being good teachers, and later, for doing the applied research that addressed social problems. For over 100 years, the land grant universities have been the leaders in applied agricultural research in the US and internationally.

Over time, however, with the changing nature of the broader society and the expansion of the university to encompass a wide range of basic sciences and professions, the land grant university system seemed to lose sight of its mission (Schuh, 1986). As a part of the expanding emphasis on basic research, the number of publications in refereed journals has increasingly become a major criterion for advancement and tenure in colleges of agriculture (Busch and Lacy, 1983), with much less credit given for more applied agricultural work such as varietal development. Thus, the status of plant breeders has fallen dramatically since the early days of land grant universities when varietal development was a central focus and plant breeders themselves were the most highly respected scientists in agriculture.

Today, the status of plant breeders has declined further within colleges of agriculture, while that of basic scientists, particularly molecular geneticists, has risen. This is evident in recent seed company advertisements which depict plant breeders as antiquated and tediously slow (see figure 3.3). The lower status of applied scientists is also reflected in scientific awards and recognition. Few applied scientists have received Nobel prizes. Only one plant breeder, Norman Borlaug, who developed some of the "miracle grains" of the Green Revolution, has been the recipient of the Nobel prize, albeit the Nobel Peace Prize. In contrast, many molecular biologists, beginning in 1962 with Watson and Crick, the co-discoverers of the molecular structure of DNA, have received Nobel prizes. Indeed, by the late 1960s, molecular biologists began to receive a surprisingly large proportion of the Nobel prizes in physiology and medicine, conferring additional status on this approach. The prestige associated with such status occasionally led to gross exag-

Some plant breeders are more patient than ours at United AgriSeeds.

Waiting for Mother Nature to reveal her secrets can be a very frustrating experience. The biggest problem is the time it takes to make advancements with traditional plant breeding.

But, there is a new family of innovative companies that is working to cut that time with biotechnology. Now **Lynks, Keltgen, Hofler** and **DeWine** are part of United AgriSeeds aggressive research and marketing organization — already the 10th largest agricultural seed company in the United States. In addition, Eurosemences S.A., in France, has established corn research in Europe.

United AgriSeeds is growing new plant lines from cells, in the laboratory, as part of our initiative to drastically reduce the time it takes to develop superior corn hybrids and soybean and wheat varieties.

Expect to hear good news soon from our impatient group of research people. And marketing people. And business people.

Our contribution is just beginning.

United AgriSeeds
P.O. Box 4011 • Champaign, IL 61820
(217) 373-5300

Harvesting the new crop genetics

Figure 3.3 Biotechnology is suggested as a replacement for antiquated plant breeding. © United Agriseeds. Reproduced by permission.

geration. For example, Watson and Tooze (1981) maintained that, for young biological scientists, molecular biology was the only field worth entering. As Yoxen (1983:43) points out, molecular biology "acquired a reputation as a field in biology where the real excitement lay and where the most daunting problems would continue to be found."

These changing research emphases can also be seen in employment patterns. In the public sector, employment opportunities in agricultural molecular genetics are growing, while employment in plant breeding is declining. Between 1982 and 1988, there was an increase from 273 to 682 full-time equivalent (FTE) scientists conducting agricultural biotechnology research in the state agricultural experiment stations. In addition, FTE staff working in this area rose from 472 in 1982 to 1131 in 1988. At the same time, overall FTE faculty positions in the state experiment stations increased by only 65 to a total of 6128 (National Association of State Universities an Land-Grant Colleges [NASULGC], 1989). Consequently, much of the increase in FTEs for biotechnology research represented a redirection of existing faculty and reallocation of faculty into this new area. Interviews with state agricultural experiment station directors indicated that many of these positions had been created by reducing the scope of conventional breeding programs. As plant breeders retire, many have been and will continue to be replaced by molecular geneticists. Between 1982 and 1986, the number of FTE plant breeders dropped from 451 to 298, with a slight recovery to 328 in 1988. Finally, the state experiment stations continue to project increases in faculty, students and staff in the area of biotechnology. Forty-one stations project 136 new faculty FTEs in biotechnology within the next two years. When compared with the actual increase experienced in 1986 and 1988, previous projections have proven to be conservative (NASULGC, 1989).

A similar process is occurring within the Agricultural Research Service (ARS). The six-year plan for 1984–90 called for ARS scientists to de-emphasize varietal breeding, to discontinue release of finished varieties, and to increasingly emphasize basic research (i.e., molecular genetics) and germplasm enhancement (USDA, 1983). In fact, the plan proposed a reduction of $2 million for breeding of horticultural crops.

A survey of public plant breeding programs (in ARS and SAESs) for horticultural and sugar crops revealed a projected 22–50 percent reduction in the number of programs and a 13–39 percent reduction in FTE positions by 1990 (Brooks and Vest, 1985). A similar survey of public breeding programs for field crops, conducted by the authors, indicated a projected 9–29 percent reduction in FTEs by 1990. Many scientists are concerned that the precipitous decline in public sector plant breeding will have dire consequences for agriculture. One scientist put it as follows: "Personnel that really know the crops as whole plants are not being replaced. These people are the ones that understand germplasm. There is a lack of recognition of the fact that plant breeding is more than making crosses and selecting. It is pivotal to the flow of information from basic to applied research" (interview). In short, the reduction in plant breeding efforts and the concomitant rise in both plant molecular biology and the use of biotechnology represent a shift in the kind of knowledge claims made by public sector scientists.

Conclusions

Within the context of our "economic" model of science, we can draw the broad outlines of plant breeding and molecular genetics as follows. Plant breeding grew out of the desires of practitioners for plant improvement. It has been practiced primarily by individuals with farm backgrounds working in land grant universities. Over a period of a century, plant breeding gradually emerged out of the practices of farmers, drawing more and more on chemistry and biology for its fundamental premises. While it has been largely driven by the demands of farmers and (later) agribusinesses, it has also supplied farmers with knowledge, strategies, and products that they could hardly have demanded themselves (e.g., hybrid corn).

Molecular biology, and subsequently molecular genetics, emerged out of the desires of Rockefeller Foundation officials to rationalize the human condition and to put medicine on what they considered to be a firm scientific footing. In addition, it reflected the convergent desires of some "basic" biologists and physicists to do for biology what had been done for physics more than a century before: to change its focus from description to control, from whole organisms to the bits and pieces that would constitute the building blocks of life. The two fields, molecular genetics and plant breeding, only began to interact with each other when, within the last decade, it became apparent that the tools developed by molecular geneticists could be used in the process of plant improvement. This, in turn, drew new actors into the system of science, including pharmaceutical and chemical companies and their recently acquired seed company subsidiaries.

It would be wrong to conclude from this that molecular genetics is somehow further removed from demand than plant breeding. It is true that the early proponents of a molecular approach to biology were diverse, but the same might be said of the early proponents of plant breeding. Only a few wealthy farmers, in alliance with farm journalists, fertilizer suppliers, urban businessmen, and bankers, supported the creation of public agricultural research (including plant breeding). The same might be said of molecular genetics. Only a few biologists, but some physicians and many businessmen, supported the creation of public molecular genetic research.

Moreover, in both cases, the strategy employed by supporters was to seek public (or foundation) financing to employ scientists in the public sector to engage in what was rightly perceived as a slow, tedious, and not very profitable enterprise. Only after the methods used became routine and the legal regime under which science took place was modified, did private financial support for these disciplines emerge. Only after control became not just a goal but a reality, did private direct investment in either plant breeding or plant molecular genetics for the purpose of product development appear desirable.

A key issue for us is what final product is desirable for whom and to whom. While the new biotechnologies offer many new avenues for plant improvement, they remain tools that may be used for many different purposes. In one sense plant breeding and molecular genetics today lend empirical support to Schafer's (1983) notion of finalized science. The central theories and technologies of both fields are well-established: Mendelian genetics for breeders and the central dogma for molecular geneticists. What remains is to decide to what purposes these finalized sciences will be put. Thus, our inquiry must focus, ultimately, on questions of ethics. However, it is first useful to look at what history can tell us about past work in plant improvement. To do this requires a closer examination of specific crop plants and the socioeconomic contexts in which they are grown. In chapters 4 and 5 we consider two crops of worldwide importance, wheat and tomatoes, and in chapter 6 we examine the special problems that biotechnology poses for international relations.

4

The Political Biology of Wheat

Wheat is one of the world's oldest cultivated plants. Its origins can be traced to the fertile crescent where it may have been gathered as early as 15,000 BC (Biggle, 1980). By Roman times, mass selection of wheat varieties was practiced. At least in large cities, bakeries began to flourish as is evidenced by the ruins at Pompeii. Although bakery production of wheat breads declined during the early Middle Ages, by 1155 London had a brotherhood of bakers supplying bread to its townspeople (Spicer, 1975).

In this chapter we first trace the history of wheat production by examining the development of a world wheat market during the last quarter of the nineteenth century. Then we examine changes in wheat production, storage, and transportation techniques. Next, we describe the government programs that have shaped the wheat market, and also note the concomitant changes that have occurred in the milling and baking industries. We then focus on how the objectives of the art and science of wheat breeding have changed in response to the changing structure of the global market. Finally, we examine the limited role played by the new biotechnologies. Throughout, it should be remembered that "the wheat industry exists as a separately defined entity only in a conceptual sense. In actual practice, practically all participants in the system are engaged in other operations, crops, or related products" (Goldberg, 1968:16).

A Brief History

Wheat was first introduced into North America by the Spanish in the sixteenth century. Being particularly well-suited to the northern Great Plains, where it can thrive on as little as 200 millimeters of rain per year, wheat moved west with the population. In the United States, before the mid-nineteenth century, Pennsylvania was the most important wheat-producing state (Schmidt, 1977). Gradually, as the population shifted and transportation over the Appalachians improved, the center of wheat production also changed. In 1874, Mennonite immigrants imported the Turkey wheat cultivar to Kansas from southern Russia. While Turkey was well adapted to use on the plains, its adoption was slowed by several factors. First, there was some confusion over just what was and was not Turkey wheat. In addition, relatively little seed was imported so harvests from the initial years were often used on the farm for consumption and the following year's seed. As late as 1902, 15,000 bushels had to be imported to meet demand (Quisenberry and Reitz, 1974). In addition, millers had difficulty in milling Turkey on stone mills and paid a lower price for it. Nevertheless, it remained the leading US hard red winter wheat cultivar until 1944. Turkey and Marquis (a Canadian wheat variety) made possible the cultivation of what until then had been called the Great American Desert.

Despite the enormous acreage of wheat planted throughout the world, it was impossible to talk about *the* wheat market until the late nineteenth century. Instead, markets were regional and local, and prices varied considerably from place to place. The problem was a simple one: wheat was a relatively heavy commodity with a low price. Without railroads and steamships it was simply too expensive to transport it great distances from its point of production.

In the mid-nineteenth century, improvements in both inland and overseas transport made wheat into an internationally traded commodity. Six countries with vast areas of land suitable for wheat growing vied for dominance of the international market: the United States, Canada, Australia, Argentina, Russia, and India. Indeed, "After 1873 separate prices in different areas of wheat production converged, on the basis of expanding quantities traded and technical improvements in transport" (Friedmann, 1978:545–6). In response to the changes wrought by the new world market, the US sent Commodore Schufeldt on a global cruise in search of new markets for all American goods including wheat in 1878–80 (Hagen, 1973). In addition, Britain invested significantly in an attempt to increase Indian wheat production and to make possible shipments to the already food-deficient British Isles. Australia, Argentina, Russia, and Canada each stepped up wheat production as well. For each major exporting country, both state policies and farm structure were altered. Let us consider briefly the strategies employed in each case.

India

The British attempted to improve wheat in India as early as 1877, soon after the opening of the Suez Canal. However, they did not mount a

full-scale effort until the establishment of the Agricultural Research
Institute at Pusa in 1904. The first set of Pusa wheats were not released
for commercial distribution until 1915 (Nair, 1979). From the beginning, however,

> India's export trade in wheat was not a spontaneous growth,
> originated and organized in India. It appears rather to have been
> developed by set policy of the British and Indian governments
> with the approval and support of the financial and commercial
> interests of Great Britain and India, on the grounds of a belief that
> the interests of both countries would be best served by an ex-
> change of raw materials with manufactured products of Great
> Britain. (Wright, 1927:380)

Moreover, the British millers felt threatened by increasing imports of
American flour. India, on the other hand, could be depended upon to
provide only whole grains, not flour, for the British market. Thus, in
addition to its mercantilist goals, the development of Indian wheat
exports appeared to offer the potential to block American incursion
into British markets. Finally, Indian wheat was harvested from March
through May, when the stocks of most other countries were depleted.

For a variety of reasons, however, India did not become a major
exporting nation. First, even by the turn of the century, India had a
much higher population density than did other wheat growing areas,
and wheat was needed to feed its rapidly growing population. Moreover, despite the high levels of irrigation, India was unable to maintain
a steady level of production. Therefore, exports were at best erratic.

Second, although much money was expended on the development of
the Indian railway system – it should be remembered that even Marx
saw the railways as the key to Indian development – the Punjab, where
most of the surplus was produced, was not connected to the port of
Karachi until 1878. As late as the 1880s, many of the best wheat
growing districts still lacked railways and the rail system was of limited
capacity. As one contemporary observer put it, "The situation had
not greatly improved by 1898, when there were few branches to the
railroads, the country roads were poor, and freights were high" (Dontlinger, 1908:230). Export was further limited by the small size of
shipments and the lack of bulk-handling facilities either at the railheads
or in the ports. Indeed, as late as 1920, India had only one grain
elevator (Wright, 1927).

A third factor limiting Indian production was Britain's emphasis
on a breeding program of soft wheat. Wright (1927:372) explains:
"Apparently in framing this policy, it [the British government] thought
only of encouraging the growth of soft wheats then preferred by Great
Britain and paid little or no attention to the possible advantages of
other kinds of wheat more suitable to other countries or to the domestic

trade". Ironically, these soft wheats were developed just as roller milling, for which hard wheats were more suitable, was becoming the preferred milling mode. However, the British milling industry resisted roller milling longer than did its continental neighbors.

Still a fourth factor involved problems in milling Indian wheat with the British stone mills, leading to its reduced price in British ports. Indian wheat flour was more often used by biscuit makers and by individuals cooking in the home, who were less concerned about yield than were commercial bakers.

Finally, India lacked an adequate system of inspection and grading – adulterated wheat (containing large quantities of foreign matter) was all too common a problem. This, combined with shipping problems during World War I, made India at best an unreliable supplier to Europe.

The United States

The history of the United States is in large part a history of westward expansion as the Great Plains were brought under the plow and the native population was killed, moved further west, or settled on reservations. The westward movement was due in part to the American tendency to farm until the soil was exhausted, and then pick up and move west again. Indeed, Maine and Massachusetts experienced declining wheat production as early as the 1830s and tried to halt it by offering "bounties" (Sutton, 1977).

The year 1862 was a watershed year in American history, not only because of the Civil War that tore the country asunder, but because of the impact the war had on future industrial and agricultural policy. In that same year four key acts were made into law by Congress, none of which would have passed were it not for the secession of the southern states. The Homestead Act opened up vast new territories owned by the federal government to settlers, many of whom planted wheat in the dry plains of the continent. At the same time, the Pacific Railway Act was passed, giving away land encompassing an area larger than England and France together in return for the construction of railways from the more populated eastern states to the west coast. The Morrill Land Grant Act was also passed, providing funds and land endowments – hence the term land grant – to each of the states for establishing colleges to teach agriculture and the mechanical arts. Finally, the forerunner of the USDA was established as a branch of the Patent Office with the power to collect statistics and to perform research. As noted above, all four acts eventually contributed to lower prices for wheat and made the US more competitive in world markets.

At the farm level, changes occurred as well. Before the last quarter of the nineteenth century, wheat farming was a capitalist enterprise in the

US. Huge capitalist "bonanza farms" utilized enormous, horse-drawn equipment. In California, farms of 20,000 to 30,000 acres were common. Hundreds of immigrants were hired at harvest time to perform that work which was not yet mechanized. In the 1870s, however, major changes occurred: new markets opened which lowered the price of wheat; household producers began smaller scale farming on cheap land purchased from the western land companies and railroads which needed settlements in order to make their investments profitable; hired labor became more expensive, making it more difficult for the capitalist farms to compete.

Canada

Canada ultimately became a major exporter of wheat, although it was late in entering the world wheat market. Until 1897–98 Canadian wheat exports were of negligible importance. Two key factors in the rise of export wheat production in Canada were the development of the Great Lakes waterway and the building of the western rail lines. Both were constructed with government subsidies of land and money.

Wright and Davis (1925) divide the early history of Canadian wheat production into three periods. The first, 1870–96, was a period in which (along lines similar to the Homestead Act in the US) free public land was made available to individuals meeting certain conditions (1871), railway-building began in earnest (1879–95), and the Dominion Experimental Farms were established (1886). Railways were essential, as few settlers had entered the western provinces prior to the building of the railways. Once the railway building program became clear, "settlers [did] not hesitate to go back 30 or 40 miles from existing lines of railroad, feeling confident that within two or three years a branch line would be built in their direction" (Smith, 1908:8). Moreover, with the distribution of Red Fife in 1882–3, a suitable variety was available for growing on the plains. In addition, by 1882, the first iron bulk freighter was able to navigate the lakes. This period was also marked by the development of grading and bulk handling. After 1887, The Canadian National railroad refused to accept grain that had not been sent through the elevators, putting farmers at a considerable disadvantage with respect to the elevator operators.

The second period (1897–1914) was largely one of rising prices until 1911–13. As a result of farmer pressure, grain inspectors became government officials and the grain trade as a whole became more heavily regulated after 1900. Soon after the turn of the century a single system of grading was adopted and supervised by the Canadian government. In 1907 Marquis wheat was introduced. Ironically, one of the parents of Marquis was a soft Indian wheat sent to Canada for evaluation. Ten years later Marquis accounted for nearly 90 percent of the acreage grown and had the effect of expanding significantly the limits of the wheat growing area (Canadian Department of Agriculture, 1967). Marquis became so important that one observer was led to state that "the prosperity of Western Canada may thus be said to depend in considerable part upon the patient efforts of generations of Indian peasants to select a quickly ripening wheat. Perhaps some new Canadian discovery will one day repay this debt" (Wright, 1927:373).

The third period included World War I (1914–25) and its aftermath. At this time demand was high due to shortfalls in Europe. Moreover, after 1914 the railroads began to overbuild. From 5,215 miles in 1905, the track mileage in the western provinces soared to 12,999 miles in 1915 (Wright and Davis, 1925). Between 1917 and 1922, most of the railroads rapidly went bankrupt and were taken over by the government.

Today, Canada exports fully two-thirds of its wheat production. This is produced on about 100,000 farms of 350–400 hectares each. All wheat is commercialized through the Canadian Wheat Board which, with the aid of a central computer, tracks all ships and freight cars (Charvet, 1988). Transport to both eastern and western ports is subsidized by the Canadian government. Moreover, despite official adherence to a fair trade policy, Canadian wheat benefits from a subsidy amounting to $73 (US) per ton (Charvet, 1988).

Australia

Australia developed more slowly than either the United States or Canada with respect to wheat production, although exports started from South Australia in the 1880s (Davis, 1940). However, in addition to the enormous distance to markets, numerous other factors weighed against Australian wheat exports. Unlike the situation in the United States, nearly all the Australian railroads were state owned and conservative in their policies. Opposition from the railroads and shippers prevented bulk handling until 1930. Private capital was not attracted to elevator construction except at mills. Most shippers, in contrast, saw bulk handling as merely increasing competition and requiring very high investments for very meager returns. Short harvest periods and variability in production also argued against development of a bulk system. New South Wales established a state-owned elevator system shortly after World War I, while Western Australia established cooperative elevators in the 1930s.

Like Canada, Australia has a Wheat Board which has a monopoly on collection and commerce in wheat. Moreover, though Australia is at the very limit of aridity in wheat production, subsidized phosphate fertilizers have made it possible to grow wheat on very poor soils. Today, 80–5 percent of Australian wheat is exported, although Australia's low population and distance from buyers still make it difficult for ships to find cargo for the return trip (Charvet, 1988).

Russia

Russia entered the world market for wheat early, exporting 1.5 million quintals as early as the 1820s. Yet, ultimately, Russia failed to capture a significant share of the market. Consider some of the factors involved.

Land tenure in prerevolutionary Russia was characterized by the absence of medium-size holdings. There were vast estates owned by the Russian aristocracy and there were small peasant holdings. After the reforms of 1861, average peasant holdings actually declined as a result of growing population. Moreover, peasant yields were generally lower than those on the larger estates because of poor quality tillage equipment and inferior seeds. The last two decades of the nineteenth century were marked by a slow growth of new cropping areas; the years just before the revolution saw much faster growth. Ironically, the land reform that occurred just after the revolution, by dividing up the land of the large estates and the wealthier peasants, created holdings that were more equal in size but often too small to be viable (Timoshenko, 1972). Nevertheless, during the last years of the monarchy, grain production, especially production of wheat, moved more toward a commercial export market. Wheat acreage increased nearly 32 percent between 1901–5 and 1913.

One factor that was of considerable importance, as it was elsewhere, was the building of railroads. The first wave of railway building in Russia was from 1869 to 1878. A second boom followed from 1894 to 1903. By the eve of the revolution, Russia possessed the second largest railroad system in the world, second only to that of the United States. However, unlike the United States, two-thirds of Russian railroads were state properties, and freight rates were controlled after the 1880s. Rate control was designed to ensure low-priced domestic supplies and to equalize the advantages of the various exporting regions. Moreover, export rates were higher than internal ones. There were no national prices for wheat (Smith, 1908). And, probably most seriously, rail rates in Russia remained very high as they declined elsewhere (Dontlinger, 1908).

Moreover, due to a lack of private capital, a system of elevators did not develop. As one observer described it, "Grain (in sacks) is usually transported from peasant farms to railroad stations or waterways in carts drawn by horses or oxen (often camels in central Asia). The distances are long, the conditions of roads bad" (Timoshenko,

1972:326). Only on the eve of the revolution did the Central State Bank begin to consider the possibility of financing a system of elevators.

Machinery import and production for use on the large estates did grow rapidly in the last quarter of the century. Machinery production doubled from 1876 to 1890, doubled again from 1890 to 1900, and increased 500 percent from 1900 to 1911. Machinery import grew even more rapidly. Two-thirds of the harvesting equipment was imported, mainly from the United States.

However, machinery, by itself, was not enough. Russian yields were low and virtually no wheat improvement work was done until after the agrarian reform of 1907. At that time the study of wheat classification began, and the agricultural experiment stations, many organized just a few years before, began to engage in systematic selection of improved wheats. Not surprisingly, results were not available before the outbreak of World War I, and yields were lowest of all the major producers.

Quality, too, remained a serious problem for the Russians. Russian wheats were often marketed as mixtures of several kinds. Because there was no official inspection service, claims of adulteration were frequent. This may have been due, in part, to the relatively small size of domestic flour mills. Even the commercial mills were small by world standards. And, as late as 1908, there were 141,000 country mills.

Despite these obstacles, the Russian presence on European grain markets was substantial. When the percentages of the market held by the United States and Russia for this period are compared, that becomes apparent:

	1888–92	1892–1902	1908–12
Russia	31%	22%	29%
United States	26%	40%	15%

It was not clear to anyone at that time that the United States would eventually occupy such a strong position and the Soviet Union such a weak one. However, the combined effects of World War I and the forced collectivization of Soviet agriculture eliminated any hopes of Russian exports of wheat. The war diverted the population away from agriculture and stopped exports entirely, while the redistribution of the land among the peasants right after the revolution made much of the farm machinery useless. It took many years before Soviet agriculture recovered.

In short, although the Russians were among the first to take advantage of the strong Western European demand for wheat, they were still saddled by an archaic monarchy, a peasant agriculture, and a small bourgeoisie that was unable to raise the capital necessary to improve agriculture. The First World War and the revolution ended any hopes of significant sustained exports.

Argentina

It can be said that the Argentinian railway system was built on wheat. In 1903, 22 of the 26 railways completed had been built to transport wheat. Nevertheless, the Argentinians were slow to adopt bulk handling, preferring instead to leave their grain in sacks that could be transported easily on ox carts. Contemporary observers complained that the wheat was often left exposed to the weather and was in poor condition when it reached the port of export. Labor was often in short supply at harvest time. In addition, one contemporary observer noted that only one-third of Argentinian wheat farmers owned their own land (Smith, 1908).

Today, much of the farmwork on Argentine wheat farms is done by "contratistas" who often do all of the work for landowners. Yet, the postharvest portion of the Argentine wheat system remains weak. In 1985, only 28 percent of grain exports arrived at the ports by rail, due to the poor quality of the rail system (Charvet, 1988). Truck transport was used instead, but its high costs kept down growers' profits.

Until recently, La Junta Nacional de Granos had a monopoly on grain exports. Moreover, those exports were taxed. On the other hand, until 1977, wheat producers could borrow at a nonindexed interest rate; given Argentina's high inflation, this amounted to a considerable subsidy. Recently, however, both the subsidized interest rates and the export taxes were eliminated, and the high rate of inflation has helped Argentina to remain internationally competitive. Ironically, Argentina's chaotic economy has made it (arguably) the sole free trader in the international wheat market.

The current international market

In sum, each of the players implemented policies, wisely or unwisely, that it hoped would encourage greater production and export. The policies ranged from improved inland transport to the establishment of research institutions. In both the United States and Canada, infrastructure was built rapidly to encourage further wheat production. Australia and Argentina, already handicapped by their greater distance to European markets, failed to develop their infrastructures as rapidly and had some difficulty in competing with the northern producers. British efforts in India met with some success, but a century of neglect of food crop research and a burgeoning local population soon put India out of the picture. Russia had the advantage of closeness to the markets, but it was saddled with an archaic monarchy and a fickle climate. Ultimately, the United States and Canada prevailed.

One consequence of the internationalization of the wheat market was the decline of wheat prices in Europe. The British took no action while the French, Germans, and Italians imposed tariffs. The Netherlands, Belgium, and Denmark did not impose tariffs but they did stop trying to export wheat (Kindleberger, 1951). Today, the world wheat trade is far larger than anyone could possibly have imagined at the turn of the century.

Entry into the international market was more than a simple function of yield. In 1900 yields were as follows (Dontlinger, 1908):

Country	Yield (Bushels per acre)
United States	13.3
United Kingdom	32.6
France	18.1
Russia	8.0
Germany	27.9
India	10.7

Clearly, the US and Russia lagged far behind other producers at that time. Yields in the United States and the average yields of other countries attained near equivalence by the 1930s. Then, the United States pulled ahead, reaching a new plateau in 1974 (Heid, 1979). Between 1970 and the present, other producers narrowed the gap. Moreover, from the early 1960s to the early 1980s world wheat trade doubled. Demand by non-wheat-producing countries increased as a result of newfound oil wealth, rural to urban migration, declining per capita food production (especially in Africa), and increased levels of food aid (USDA, 1984b). Today, 75 percent of world wheat production enters the market and 29 percent is traded internationally (CIMMYT, 1985), compared with a mere 4 percent for rice. The United States, Canada, and Australia account for three-fourths of the world's wheat exports, although China is now the world's largest producer. While United States exports are entirely in the private sector, Canada and Australia both have wheat marketing boards that control all export sales of grain, and since 1979, Australia has underwritten the price of wheat (Industries Assistance Commission [IAC], 1988). Argentina has a National Grain Board but it does not enjoy monopoly status.

Government Programs

State intervention in wheat production, processing, sale, export, and research is both ubiquitous and traditional. Ancient Rome set policies for wheat production and baking. In recent times, too, the state has played an important role. Indeed, as early as 1908, Peter Dontlinger (1908:31) could write that "the National Governments of all the principal wheat growing countries of the world are factors in an official capacity in the culture of wheat, and at times millions of dollars are expended by a single government in endeavoring to solve some single problem of unusual importance." Nor did state intervention cease after the establishment of world markets.

Today, there appear to be numerous reasons for state intervention in domestic wheat markets.[1] First, there is a concern with stability. Stability of prices reduces the farmer's risk and encourages more intensive crop production. Second, states are concerned with the problem of food security. No nation wishes to be dependent on others for staple foods. Third, virtually all states desire a cheap food policy, either from humanitarian concerns for the poor or from a desire to avoid civil unrest, or both. Fourth, governments often intervene as a way of saving or creating foreign exchange. For major wheat exporters, wheat is an important earner of foreign exchange; for importers it can be a significant drain on the economy. Fifth, millers, bakers, and consumers often demand protection against adulterated products, which usually involves inspecting and regulating wheat, flour, and bread. Other goals include conservation, agricultural development, and "aiding the free market".

A variety of policies are used to reach these goals, including negotiating prices among various interest groups, indexing based on inflation or on world prices, allotments based on acreage or quantity, maintaining buffer stocks, and subsidizing research. All of these policies affect returns to producers, costs to users, and the quality and availability of wheat products. And, ironically, these very policies may serve to destabilize the world market (Zwart and Meilke, 1979). Of course, even a cursory examination of all the policies of the major wheat producing and consuming nations is beyond the scope of this volume. However,

---

[1] For a more detailed review of some of the reasons why states intervene in wheat markets, see CIMMYT (1985) and Charvet (1988). The Industries Assistance Commission of Australia estimates that "government supported agricultural interventions are estimated to have caused the world price of wheat to be around 15 per cent lower than otherwise (with the policies of the EEC and the US being the most influential)" (IAC, 1988:65). Of course, while this may be accurate, there are powerful interests, including some farmers, for whom such low prices are desirable.

given the importance of the United States to the political economy of world wheat production and trade, a review of American wheat programs at the farm level is in order.

Despite low prices and the resultant periodic collapse of the farm economy, the US government avoided direct intervention in the wheat market until well into this century. Ironically, the first intervention, in 1916, resulted from temporary shortages brought on by the disruption of production in Europe during the World War I and by the massive increase in demand for US wheat. Temporary export controls and price limitations were imposed. As soon as the war ended, however, low prices once again became the order of the day. During the 1920s – a period of extended farm depression in the United States – pressure mounted to develop a price support system based on parity, the average of the prices paid during 1910–14, when the war demand was highest.

The Republican administrations of the 1920s refused to deal with the issue. When they finally did intervene, the Agricultural Marketing Act of 1929 did too little too late. The act established a marketing corporation to make loans to marketing cooperatives that would, in turn, purchase wheat from farmers. By 1930–1, it accumulated stocks of 313 million bushels of wheat. It soon ran out of money and ceased to function (Hadwiger, 1970; USDA, 1984b).

The programs of the Roosevelt era bore the imprimatur of farmer, seedsman, journalist, and Secretary of Agriculture Henry Wallace. Wallace envisioned a world economy in which planning would ensure that the enormous abundance of wheat was properly and fully utilized. In 1933 he negotiated the International Wheat Agreement by which wheat exporting nations agreed to a 15 percent reduction in production. In the same year the Agricultural Adjustment Act assigned wheat acreages to US farmers based on their past production. However, by virtue of the program design, recipients paid under the program were largely one-crop farmers who continued to plant that same crop (Worster, 1979). Given the complexities of the programs, legislators conceived of a model wheat farm when they drafted the act. In practice, this meant that the rules were designed for the convenience of the larger, specialized producer. This, in turn, affected participation. Hadwiger (1970:317) explains:

> In practice the medium- and large-sized farms have received most of the direct benefits from wheat commodity programs. Program rules have been framed for the convenience of the larger farmer. The acreage cuts have been inherently unacceptable to the farmer whose total farm size was already inadequate. For these and other reasons fewer small farmers have participated in the programs and the benefits flowing from them.

This domestic allotment program remained in place through the drought years of 1933–6. However, in 1936, the Supreme Court ruled that the Agricultural Adjustment Act was unconstitutional and the program was disbanded.

The Soil Conservation and Domestic Allotment Act of 1936 stressed soil conservation as a response to the severe soil erosion problems which resulted in what quickly came to be known as the Dust Bowl. The Agricultural Adjustment Act of 1938 established a system of loans for farmers in which eligibility was tied to an agreement to abide by acreage allotments. Under the terms of this act, farmers were provided with federal price guarantees whenever prices fell below parity. The law also provided that quotas could be imposed if supplies exceeded 135 percent of demand. The American Farm Bureau Federation, at the zenith of its power at this time, was the major supporter of this legislation.

World War II saw the establishment of voluntary controls in 1939–40 and compulsory controls in 1941–2. Surpluses began to accumulate immediately and exceeded the capacity of all storage facilities by 1942. The USDA tried to eliminate most of these surplus stocks over the next few years, and was so successful that when they were desperately needed in 1946 to feed a hungry Europe, they were no longer there. As Hadwiger (1970:33) put it: "The official USDA policy favored a 'bare shelf' at the end of the war. The USDA took the view that postwar food relief needs would be small and probably could be met from army food stocks on hand at the time." By the winter of 1945–6, a crisis had developed because of European food needs.

The late 1940s were marked by rising wheat production. The Agriculture Acts of 1948 and 1949 revised the parity formula for loans but made few changes in the overall concept of the program. The Truman administration continued the high price supports and reinstituted the stockpiling of large surpluses. By 1955, wheat stocks were estimated at 996 million bushels. The Soil Bank was established in 1956 to withdraw farmland from production. Nevertheless, by the 1960s there was an immense wheat surplus.

During the 1960s a series of acts were passed to reduce the mounting wheat surplus. The Agricultural Act of 1964 reduced the loan rate and made participation in the program voluntary once again. The following year, the Food and Agriculture Act of 1965 extended that voluntary program and added a cropland adjustment program to reduce acreage further. However, significant reductions in the surplus awaited the authorization of massive international food distributions under the terms of Public Law 480, the Agricultural Trade Development and Assistance Act of 1954 (10 July 1954, ch. 496; 68 Stat. 454).

Food aid, under the 1954 act, seemed the solution to both the problems of the Cold War and domestic overproduction. At the time it was felt that the rapid path to development was immediate industrialization. Thus, the aid was directed at feeding industrial workers, not at improving Third World agriculture. In fact, the low-cost or free wheat depressed prices and discouraged production of locally grown grains beyond what was immediately needed for peasant consumption. It also had the effect of turning relatively self-sufficient agricultural societies into urban ones dependent on the world market for food. Friedmann (1982:S269) explains: "Most countries attempting to industrialize adopted cheap food policies. These were made possible by cheap imports with minimum drain on foreign exchange. The policy was economically rational within the international food order and politically pressing as population growth outstripped industrial employment."

This system only worked, however, as long as the United States had huge surpluses to give away or to sell below market prices. By the early 1970s the world economy had changed such that (1) the United States could no longer afford to give the grain away, (2) the Soviet Union proved to be a far more profitable market, and (3) the Third World nations could no longer afford to buy at the new, higher prices.

The Agricultural Trade Development and Assistance Act of 1954 now benefited the white-wheat producers of the Pacific Northwest because they were furthest from domestic markets and their soft, white wheat was preferred in India. Indeed, wheat has consistently replaced other grains on the world market; that is, one effect of Public Law 480 was to increase the number of wheat eaters in the world, in part, at the expense of other grains. Moreover, for the millions in the Third World whose ties to the land have been severed, food is now a commodity, only purchasable with cash earned from scarce industrial jobs. These persons are often tied to wheat consumption as no other grains are available in sufficient quantities to supply the demand. In Africa, for example, wheat imports doubled between 1975 and 1985 (CIMMYT, 1985). Therefore, as income rises in Third World nations, consumers buy more bread; as it declines they switch to buying flour. "Misery in the Third World and recession in the First are now intimately connected" (Friedmann, 1982:S283).

The Agriculture Act of 1970 first limited the maximum payment to any one farmer to $55,000 per annum per crop. A major shift came with the Agriculture and Consumer Protection Act of 1973 which moved to a market-oriented program. A target price was established with the maximum payment equal to the difference between the target and loan rates. The goal of the act was explicitly to support income without affecting the market price. The Food and Agriculture Act of 1981 continued the target price system and also authorized crop-specific acreage reductions. The payment-in-kind program of 1982–3 was designed to empty the overflowing American granaries. The idea was fundamentally a simple one: farmers would be provided with warehouse receipts for grain they could either sell or use as they saw fit. In return, they would agree not to plant the crop in question. In 1982–3 there was a 15 percent acreage reduction, a 5 percent cash diversion, and a 10–30 percent payment in kind. The high level of participation in the payment-in-kind program was attributable to the elimination of the payment limit first instituted in 1970. Indeed, 5 percent of farmers with 1,500 acres or more received 27 percent of the total payments (USDA, 1984b).

In short, the US wheat programs have varied considerably over the years, in part because the situation was not always fully understood, in part because a lack of timely response, and in part because of a lack of political responsiveness to the needs of (all) farmers. Until recently, these programs have been largely *ad hoc* and lacking in comprehensiveness. Moreover, despite the professed interest in a variety of often lofty goals, "under every major program the income objective has invariably become preeminent even when not featured in the law" (Hadwiger, 1970:382).

Production

World wheat production today encompasses virtually every size of farm from the smallest to the largest. However, it is the largest producers who produce for the world market. In the United States, for example, wheat is the third leading crop in terms of value. Figure 4.1 is a schematic diagram of the industry. According to the 1978 Census of Agriculture, 383,000 farms harvested wheat but 65 percent of these produced the crop on less than 100 acres, which is small by American standards. Five major types of wheat are grown in the United States today: hard red winter, soft red winter, hard red spring, white, and durum; the winter wheats account for 79–80 percent of production (USDA, 1984b). Three types of land-use situations are typical of wheat in the United States: (1) continuously cropped dryland in the central and eastern plains, (2) summer fallowed land in the western plains and the northwest, and (3) irrigated fields in Kansas, Oklahoma, and Texas. Irrigated wheat, however, is dependent on the water supplied by the Ogalalla aquifer. As the aquifer is depleted and oil prices (for pumping) rise, it is likely that this irrigated farming will disappear. In short, wheat is usually grown in areas where there are few alternatives as a result of extreme dryness and/or short growing seasons. In some areas, such as the southeastern United States, it is double cropped with soybeans.

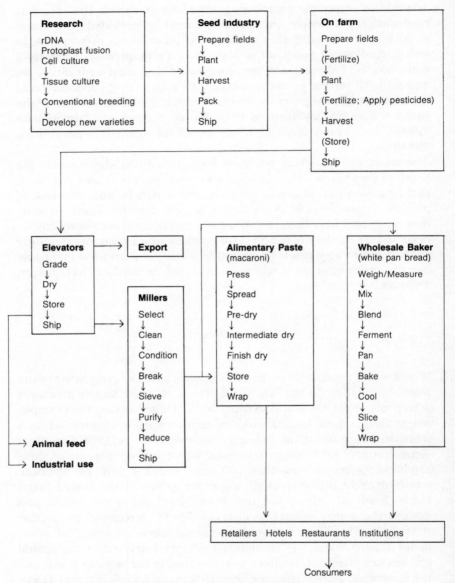

Figure 4.1   A schematic diagram of the US wheat industry.

Because wheat is often grown in places where few alternatives exist, it lends itself to a peculiar kind of one-crop economy. Its economy is unstable with most of the labor concentrated at planting and harvest. In 1950, production and harvest required only 15 hours per acre; by 1975 a higher level of production required less than 2 hours (Heid, 1979). As a result of the low labor needs and the lack of alternative on-farm employment, US wheat farmers often combine their farming with off-farm employment. By 1970, one-third of the income of wheat farmers was derived from off-farm sources (Hadwiger, 1970). Today, with depressed prices and more efficient equipment, that proportion is undoubtedly higher.

Mechanization

Wheat in the United States was in the forefront of mechanized production. The first threshing machine was developed by the Scotsman Andrew Meikle in 1784 (Splinter, 1974). In 1834, Cyrus McCormick invented his very successful reaper, borrowing from and improving upon other less successful designs. As early as 1880, four-fifths of US wheat was reaped by machine (Giedion, 1975). In California, where farms tended to be larger and more specialized, combines drawn by as many as 30 to 44 horses were in use by the 1850s. Steam tractors replaced these combines in the last years of the nineteenth century. The steam harvester had a cutting bar of 24 to 42 feet in length and could harvest as much as 75 to 125 acres per day with a crew of eight (Dontlinger, 1908). These mammoth machines cost the extraordinary sum of $7,500 at the turn of the century. In 1903, 3,000 steam harvesters were in operation on the Pacific Coast. Harvesters were restricted to use in the West until 1922 when the gasoline tractor power take-off was developed, permitting mechanized harvest of smaller fields. Self-propelled combines are typical today, and can even be found in the hilly regions of Washington and Idaho. Machinery has been so important in wheat cultivation that one observer remarked that, "more than any other crop, wheat has attracted the interest of the inventor and the innovator through the centuries. Most technological developments have moved from wheat production technologies to other cropping systems" (Splinter, 1974:69).

While early wheat varieties were tall-strawed and produced uneven stands, the introduction of machinery necessitated the development of machine-harvestable varieties. Moreover, as machine harvesting was rougher on the wheat plant, it posed new problems for the wheat breeder (Alsberg and Griffing, 1928). In particular, resistance to lodging (falling over) was essential and could be accomplished either through (1) breeding varieties with stiffer straw, (2) making the straw more pliable so that it would bend in the wind, (3) increasing the root strength of the plant, or (4) dwarfing. Initial attention was paid to the development of stiff straw, while more recently the introduction of semi-dwarf varieties from Japan and Korea after World War II has been particularly significant. With the introduction of machines, resistance to shattering was also important. The fact that the machines hit the wheat harder than a sickle or scythe meant that individual grains were much more likely to fall off the plant. Resistance to shattering, therefore, became a breeding goal. As machinery has increased in size beyond the ability of farmers to own their own equipment, custom harvesting operations have become common. However, this development has intensified the need for resistance to shattering and lodging, as wheat must remain in the field until the harvesters arrive and is consequently subject to wind and rain damage.

Another effect of mechanization has been the reduction in labor needed to grow and harvest wheat. While detailed crop-specific data are hard to come by, some figures make the point. At the turn of the century, 28,000 workers were needed in Kansas alone to harvest the crop (Dontlinger, 1908). In addition to reducing the total labor needed, mechanization has shifted the labor use to the processing sector. In 1959, a total of 524,000 persons per year were needed for the entire US wheat industry. Of these, 109,000 were in production, 55,000 in input supplies, and 360,000 in processing (Hadwiger, 1970).

Finally, the use of large equipment has meant more rapid harvesting of wheat crops. This has increased the magnitude of the peaks and troughs in the demand for transportation and storage. It is to the storage and transportation system that we now turn.

Storage and Transportation

Wheat has been stored and transported in the United States as a bulk commodity for about a century. At the turn of the century an investigation by the Interstate Commerce Commission found that combinations (trusts) between the railroads and elevators had nearly eliminated price competition. This, in turn, brought on farmer pressure for government regulation of both storage and transport prices. As early as 1848, the Chicago Board of Trade developed grain standards. However, it was not until passage of the Grain Standards Act in 1916 that the first US national standard was developed (Chicago Board of Trade, 1982).

Storage today is more than just warehousing; at the elevator the farmer's wheat is graded, cleaned, and weighed. Country elevators (where farmers deposit their grain) may also sell feed. They are either independent, farmer cooperative, or line elevators (part of a chain owned by a grain company). Typically, country elevators sell to subterminals or terminals. Federal and state licensing of elevators is available, although it is not mandatory. Licensed elevators must meet certain standards which may be desirable to both farmers and elevator operators. Of particular note is that only federally licensed elevators can store government grain. Wheat is inspected at the elevators by both federal and state agencies, though only grain for export is required to be inspected. On the basis of the inspection, the wheat is placed in one of seven classes and 13 subclasses.

Subterminals were built starting in the late 1970s to obtain better freight rates by shipping 50–100 car unit-trains of grain. Terminals receive most of their grain in these same unit-trains. The terminals also provide blending and drying services. Some of the terminals are "private", that is, owned by processors or millers, while others are "public", (i.e., open to anyone). Only public elevators are licensed by the federal government. Terminal markets stock grain of various grades all year round. Today, the four largest grain elevator companies account for 23.6 percent of the sales. Among grain agents, brokers and commission merchants, the four largest companies account for 47 percent of sales (US Bureau of the Census, 1985).

Much inland transport of wheat is still carried out by rail, but truck transport is favored by some because of ease of loading and unloading. Rail transport is regulated, although the level of regulation was reduced by the Staggers Rail Act of 1980. Rates are so complex as to lead one observer to say that "the grain rate structure in its present form has been appropriately described as a 'monster'" (Chicago Board of Trade, 1982:64). International transport of wheat is normally by sea. Today, six firms control most of the US wheat exports (Heid, 1979).

Milling[2]

Flour milling is almost as old as the planting of wheat itself. In ancient times, flour milling was an arduous, hand-labor process in which two stones or a mortar and pestle were used. In the Middle Ages, partly as a result of the desire of feudal lords to ensure that taxes were collected, wind and water mills ground flour for each locality. Initially, flour making consisted of simply grinding the entire wheat kernel into a fine powder. However, it was soon noted that if the wheat was only partial-

[2] In this section and the following one we focus, for simplicity, on the milling and baking of white pan bread. Other wheat products employ different milling and processing techniques. For example, Durum wheat is milled into coarse middlings called semolina and then processed into macaroni products.

ly ground and the finer particles then separated from the coarser ones by means of a screen that a white flour (and consequently white bread) could be produced. White flour is known to have a lower nutrient quality than whole wheat flour since the nutrients are not found in the endosperm. Nevertheless, Western traditions linking purity with whiteness prevailed, especially in North America, England, and Southern Europe. At first only the wealthy desired white flour and bread; then nearly everyone preferred it.

The milling process remained nearly the same for centuries until 1785 when Oliver Evans invented a prototype of the modern flour mill. Evans's idea was a simple one: he developed a system whereby the wheat was carried to the top of the mill and the flour was separated from the bran automatically with little or no hand labor. He also wrote a manual for millers, providing detailed instructions for reproducing his design. Evans's design was so vastly superior to other mills of the day that by 1792 over 100 mills had been built using it (Ferguson, 1980). Although the power source changed from water and wind to steam over the years, Evans's design remained unchanged until the 1870s when roller and burr mills were developed.[3] Evans's mill employed a process known as "low-grinding" in which the object was to separate as much flour as possible from the bran in the first pass. In contrast, roller and burr mills break the kernels in successive grindings (Mahlin, 1973). The development of roller mills increased interest considerably in American, Canadian, Argentinian, and Australian hard wheats. In fact, roller mills were developed in part as a response to the hardness of Hungarian wheats. The new milling method sparked such interest in the United States that a group of Minneapolis millers went to Hungary in 1873 to learn the new process (Quisenberry and Reitz, 1974). Roller milling also encouraged the development of larger mills since economies of scale were greater (Alsberg and Griffing, 1928).

Yield for the miller is measured in flour per unit of wheat. However, the process of flour production is by no means simple. "The shape and the structure of the wheat grain are responsible for the complexity of the milling process, which seeks to obtain the maximum yield of saleable white flour with the minimum bran content" (Eggitt et al., 1975:219). Moreover, a large crease in the wheat germ reduces the potential level of extraction that can be achieved without contamination from the germ or bran. The exact yield a miller achieves may vary depending on the relative prices of flour and mill offals (i.e., bran and germ). For example, if the market price for mill offals is high, for use as

---

[3] Roller mills, originally developed to process sugar cane and roll metal, were soon found to be well-adapted to flour milling (Chicago Board of Trade, 1982).

animal feed or for other products, then the miller might decide to leave more endosperm in the offal. This means, in turn, that the flour will be of a higher quality because it will be whiter by virtue of not being discolored by the germ and the bran.

Hardness is also related to yield for the miller. Hardness is correlated with the fracturing of the endosperm in wheat. Hard wheat tends to break into regular particles while soft wheat produces irregular ones. A critical level of "starch damage" (i.e., breaking of starch granules) is necessary in milling for successful bread making. Moreover, in hard wheat the endosperm separates more easily from the germ and the bran, permitting higher flour yields (Kingswood, 1975).

Changes in flour milling technologies have also affected the size and number of flour mills. The flour milling industry in most nations is quite concentrated. In the United States, the number of milling companies declined from 1,084 in 1947 to 251 in 1982. At the same time, the four-firm concentration ratio (i.e., the market share of the four largest millers) increased from 29 percent in 1947 to 40 percent in 1982 (US Bureau of the Census, 1986). In Britain, concentration occurred even earlier; by 1935 the three largest millers accounted for 39 percent of the flour output (Maunder, 1970?). In Australia, concentration has occurred more recently and more rapidly; three companies now control over one-third of the mills and one-half of the milling capacity (IAC, 1978). Ironically, however, the largest and most modern flour mills today are not located in the United States or other Western nations but in Nigeria, Indonesia, and Sri Lanka where huge import shipments are routinely processed into flour in the port cities for distribution throughout the respective countries. These large mills can capitalize on low local labor costs as well as efficient machinery, even making possible re-export of flour to surrounding nations (CIMMYT, 1985).

The shift in the geographic location of wheat production at the turn of the century also affected the milling industry. While Budapest was the world's largest milling city until 1890, it was surpassed soon after by Minneapolis (Dontlinger, 1908).

Bread Making

Baking has long been a specialized art. In the early stages of the industrial revolution, the mechanization of baking played a significant role, although it should be remembered that, at least in the United States, most flour was sold directly to consumers until after World War II. In contrast, by the 1970s 95 percent of the flour was processed into bakery products. By 1789, dough-kneading machines were in operation in Genoa, and as early as 1833, biscuits in England were produced on an assembly line (Giedion, 1975). Continuous flow production soon became a preoccupation of the industry.

Nevertheless, although it was relatively easy to develop tunnel ovens with conveyor belts to bake the bread, dough preparation remained a laborious process. Mechanical kneading became possible in this century and was rapidly adopted. Mechanized kneading produces higher bread yields than hand kneading because of increased water absorption. However, full mechanization required not only standard flour but also standard yeast. Fermentation took both time and space in the bakery, but it was nearly impossible to dispense with, because it increased the elasticity of the dough, producing both a finer texture and greater keeping quality. "It was the discovery that dough could be developed [i.e., brought to baking readiness] by means other than yeast fermentation that eventually led to the explosion in bread-making technology which has characterized the last 20 years" (Chamberlain, 1975:262). The mechanical dough developer was first developed in the United States and was first used commercially by an individual with the improbable name of J. C. Baker in the early 1950s. This invention made continuous dough production possible for the first time. By 1975, 30–40 percent of American bread was made this way.

In contrast, the Chorleywood bread process (CBP), employed in the United Kingdom, uses different ingredients and machinery. Its central feature is automatic batch mixing rather than the continuous process used in the United States. By 1972, 75–80 percent of all British bread was manufactured in this way. This process also permitted the use of lower protein flour, a significant bonus for the baker.

Chemicals began to play a role in bread manufacturing in 1850 by whitening the dough or improving its plasticity;[4] they also occasionally reduced nutrient content (Elton, 1975). Most recently, chemicals have been used in the activated dough development process to speed dough development.

The finished product remained unsliced and unpackaged until the product was sufficiently uniform – in both size and texture – to move through a slicing machine. In the United States, bread was first wrapped in 1909 and first sliced in 1920 (Heid, 1979). The most recently introduced technology for baking has altered flour requirements such that low protein (and hence higher-yielding) wheat can be used to make products that previously required high-protein wheat. Giedion (1975:210) remarks about the changes in the milling and bread-making process: "The changed characteristics of bread always turned out to the

---

[4] Chemicals as adulterants of bread can be traced back considerably further. Indeed, the UK prohibited the adulteration of flour in 1758 (Elton, 1975).

benefit of the producer. It was as if the consumer unconsciously adapted his taste to the type of bread best suited to mass production and rapid turn-over."

Concentration in the baking industry has occurred more recently and more rapidly than in milling, perhaps in part because of the consumers' shift to purchasing bread rather than flour. In 1947, in the United States, there were 5,985 commercial bakeries, four of which controlled 16 percent of the market. By 1982, there were only 1869 bakeries and the largest four controlled 34 percent of the market (US Bureau of the Census, 1986). In contrast, in Britain in 1970, four firms accounted for 70 percent of the bread market (Maunder, 1970?).

The emphasis in Western wheat-growing countries has been and remains on bread making. Relatively little research has been done on other wheat products, especially the unleavened breads and other products of the Third World (Irvine, 1975; CIMMYT, 1985).

Breeding Wheat for Bread

Wheat may be simply divided into winter and spring wheats. Winter wheat must be planted in the fall to successfully form heads or spikes. Moreover, in much of the world, wheat is particularly well-adapted to wet winters and dry summers. Hence, the Mediterranean basin has traditionally been a wheat-growing area. In contrast, spring wheat is planted in spring and harvested that same year. It is more suitable to climates with particularly harsh winters or to areas where freeze and thaw cycles occur through the winter.

At the genetic level, wheat varieties may be diploid (one set of chromosomes), tetraploid (two sets of chromosomes), or hexaploid (three sets of chromosomes) with these three types containing 14, 28, and 42 chromosomes, respectively. The presence in hexaploid varieties of as many as six alleles (gene copies) for a single characteristic permits "fine tuning" of a number of crop characters (Riley, 1975). However, this allelic multiplicity also complicates the breeding process since a given gene locus (whether desired or not) may appear on six different chromosomes in six different forms. Moreover, the phenotypic expression of a desired characteristic may require the presence of the same allele in each of these locations. Diploid varieties, while lacking this chromosomal complexity, are useful for breeding. Because of their smaller chromosomal complement, they may successfully breed with wild relaties producing fertile tetraploid or hexaploid hybrids (Picard, 1988).

Although wheat improvement is an ancient art, probably practiced by our remote ancestors, wheat breeding is a relatively new phenomenon.

By the early nineteenth century, American plant explorers were looking worldwide for new varieties for American farmers to try. Even Commodore Perry's celebrated trip to Japan had, as part of its mission, the collection of exotic seeds. The farm journals widely publicized new varieties and, if successful, they were quickly adopted by American farmers (Gates, 1960).

By the turn of the century, the USDA had established a regular process of testing new varieties. For example, between 1896 and 1900, 245 varieties from around the world were tested in the United States. Those that proved promising were then released to farmers. The process was simple enough that one breeder could suggest a "simple method ... which if practiced, would enable any farmer to constantly and effectively improve the yield and quality of grain with little trouble, but with great profit in the end" (Carleton, 1900:68).

A key aspect of wheat breeding has always been determining its objectives. Until the development of a world market in wheat, yield was not only the primary but often the only objective. The reasons for this were simple enough: no tests or standards for quality were available, and little was known about the control of insects and pathogens. With the development of a world market, however, it became possible to compare the relative qualities of wheat grown in different places. Moreover, these qualities were not merely the apparent ones – color, shape, fullness of berries, etc. – but the baking and milling qualities as well. Good flour and bread have always been desired, though the definition of "good" varied from place to place. Yet, without an international market for wheat, no additional economic value could be placed on such qualities. Indeed, most variation was probably seasonal. In some seasons, wheat produced good flour and/or bread, and in others it did not. The reasons were unclear and, in any case, were beyond the control of farmers, millers, bakers, and consumers alike.

The development of an international market changed the breeding enterprise markedly. As one observer put it 60 years ago: "In reality, the factors determining the objectives of the wheat breeder are not merely complex, but they change with changing economic conditions and with changes in the economic factors determining farmers' profits" (Alsberg and Griffing, 1928:269). We will now consider breeding for quality.

Quality

Wheat breeders commonly assert that they breed for quality. However, this apparently simple character, in fact, glosses over many different and sometimes conflicting breeding objectives. As Blaxter argues: "Quality ... is an aggregation of attributes and to define it entails defining a purpose" (Blaxter, 1975:339). In the mid-nineteenth century, quality meant plump, large berries, and free-milling (i.e., the flour separated easily from the bran). Charles Saunders, a breeder at the turn of the century, considered seven characteristics when assessing quality: relative plumpness of kernels, density of kernels, moisture content, soundness of grain (i.e., freedom from fermentation), baking strength, absorption (i.e., the amount of water the flour would absorb), and the color of the flour (Buller, 1919).[5]

To the miller, quality often refers to the yield of flour as compared to a standard wheat, while other factors, such as amylase content and protein quality are assessed separately (Eggitt et al., 1975). A recent CIMMYT report defined quality as "the summation of several milling, baking and physical dough characters which may be conveniently considered as separate breeding characters" (CIMMYT, 1985:80–81). What all of these definitions conceal is the fact that all quality characters cannot be maximized at once. Moreover, while high protein content is desired for bread production (in the United States, 12 percent protein content is considered adequate for bread manufacture), protein is inversely correlated with yield. This is true not only across varieties but across years. A good harvest means lower average protein content than a poor harvest. Thus, to a significant extent, the desires of bakers can only be satisfied at the expense of farmers.

Initially, the dilemma for the breeder was telling whether a variety that yielded more to the farmer would also command a higher price (or at least not a lower price) on the market than other varieties in use. That price, of course, was a function of the milling and baking qualities of the wheat. Therefore, the breeder needed a way to determine what kind of bread would result from a new variety.

The case of Charles Saunders, developer of the celebrated Marquis wheat, is instructive. Saunders was appointed to the post of "Cerealist" at the Dominion Experimental Station in Ottawa in 1903. At that time, Red Fife, in use since the 1880s, was the dominant Canadian wheat variety. Saunders's father, William, had tried a promising Russian variety, Ladoga, and found that it matured earlier, solving the problem of early frost. However, an entire carload of wheat was necessary in order to test its milling and baking qualities adequately. When enough wheat was finally grown, Ladoga was found to produce a yellowish, coarse loaf.

Charles Saunders found that chewing the grain and noting its color and elasticity provided a good guide to its baking qualities. The chewing test became so important that it led one observer to remark: "It

---

[5] Note that high moisture content and unsound grain are largely problems in the farmer's field and are not resolvable through breeding programs.

requires some patience and a fairly good set of teeth, but these two attributes can be considered essential to all breeders of wheat" (Buller, 1919:202). In developing Marquis, he found maturation earliness to be as important as it had been to his father. However, "from the first, when selecting new wheats, Dr Saunders bore the requirements of the British market in mind" (Buller, 1919:200).

Marquis solved many problems for farmers, millers, and bakers. It ripened a full six days earlier than Red Fife, providing more time for harvest and land preparation before the frosts arrived. It provided more uniform grain for the millers. Most importantly, it sold well in British markets because it produced the preferred white, uniform loaves. Millers in milling companies in the US found Marquis to be so desirable that they imported it and sold it at cost to American farmers. One company, the Russell, Miller Milling Company of Minneapolis alone imported 111,000 bushels of seed wheat. The dramatic adoption rates make the story clear: In 1914, 3 percent of Minnesota farmers grew Marquis; by 1916, 30 percent and by 1917, 46 percent (Buller, 1919).

It is instructive to compare the success of Marquis with the failure of Blackhull, a variety released by the Kansas Agricultural Experiment Station in 1917 (Flora, 1986). Blackhull was well received by farmers because it yielded several bushels more per acre than other hard red winter wheats (such as Turkey and Kanred). However, millers and bakers disliked Blackhull because it yielded less flour and had a lower gluten content. It also performed less well in the new high-speed mixers developed for dough making (Alsberg and Griffing, 1928). Moreover, it was indistinguishable from other hard red winter wheats when marketed. By 1927, fearing that Kansas wheat would come into disfavor with the milling industry, an experiment station bulletin recommended that farmers grow Turkey again. Tenmarq, a variety released about 1930, was the first Kansas wheat to be bred for improved baking quality.

Determining baking quality remained a problem until the late 1920s. It had been long known that protein (i.e., gluten) causes flour to rise. Protein increases the absorptive capacity of the flour as it turns into dough, yielding more loaves per unit of flour. However, early methods for determining protein content were time-consuming and laborious. In 1883, Kjeldahl, a Danish chemist, developed a simple method to determine the nitrogen content of wheat, from which the protein content could be determined, and that, in turn, was a fairly reliable indicator of baking quality. However, breeders ignored Kjeldahl's method because it required too much seed. In the 1920s, methods were developed for estimating the protein content from just a single seed. At this point, large commercial bakeries began to hire chemists to test the protein content before flour was purchased. Today, millers, bakers, and exporters all specify a given protein content.

A more recent incident shows that it is not only the interests of millers, bakers, and farmers that enter into the breeding process, but also those of elevator operators. Given the depressed hard red winter wheat prices in the late 1970s, the Kansas Experiment Station decided to explore the development of white wheat varieties, believing this would open a new market to Kansas producers. However, elevator operators resisted this research proposal for obvious reasons: such a change would require additional elevator space and the development of new marketing channels as the white wheat would have to be stored and marketed separately from other wheat. However, the total amount of wheat marketed would not be increased, resulting in lower profits for the elevator operators. The project was soon abandoned (see Flora, 1986).

Over the years, the very success in wheat breeding has changed the research agenda. Flora (1986:201) explains:

> Increased yields created new research agendas. The heavier heads led to more lodging, and, then, a shorter, stiffer straw to resist lodging again became a consideration. Hence, the recent movement to semi-dwarf varieties. The introduction of new strains of wheat with substantially higher yields induced planting of a limited number of varieties with a resultant vulnerability to diseases and pests. Breeding then attempted to counter that vulnerability.

Undoubtedly, as the new biotechnologies begin to have an impact on wheat, those agendas will change again.

Germplasm

In the past, wheat varieties for breeding purposes were collected from other nations on plant exploration trips. More recently, improved varieties have begun to display native landraces, in part, due to the desires of farmers to grow higher-yielding varieties and to the standardization requirements of the marketplace. As early as 1928, Alsberg and Griffing (1928:285) argued that "the more perfectly a wheat-growing region can standardize its wheat, the more uniform and invariable it is, the greater the sense of security to the buyer." In short, the standardization required by the miller and baker has forced farmers into reducing the variability and the number of varieties in their fields.[6]

Consequently, germplasm banks have been established in many coun-

---

[6] In Australia, the Australian Wheat Board is empowered to discount the price of nonrecommended varieties of wheat delivered by farmers. This creates a substantial desire among farmers to conform to the approved list (IAC, 1988).

tries. Unfortunately, no single institution is the center for wheat germ-plasm conservation (Marshall and Brown, 1981). Moreover, wheat's enormous diversity of uses and the differing climatic zones in which it is grown make compiling a single descriptor list of the various collections an arduous and perhaps impossible task. Indeed, some have voiced the concern that breeders will be swamped by a great deal of irrelevant information. Marshall and Brown (1981:37) explain: "The diverse aims of breeding programs and the location-specific nature of many of the characteristics of importance in breeding, raise doubts as to the rele-vance of any list which claims to be of general or universal interest to breeders."

Crop uniformity in the field has raised some concern about the possibility of a major disease or pest outbreak. For example, a 1979 OTA study found that currently-used wheat varieties were all vulner-able to rusts (Doyle, 1985).

The passage of the Plant Variety Protection Act in 1970 permitted patent-like protection of plants for the first time. A recent report by the Council for Agricultural Science and Technology (CAST, 1985) noted that no increase in private research and development has resulted, although increased breeding of soybeans and wheat is apparent. To date, about one-third of the certificates issued have pertained to these two crops. On the other hand, the same report found that "without patent-like protection, scientific findings made during the development process have less potential monetary value and hence are more likely to be shared" (CAST, 1985:29). One private sector administrator put it this way: "[We] have had a proliferation of [wheat] varieties in the marketplace – but they are not necessarily the best varieties. Fun-damentally, the diversity in the crop is massive – the question is how to utilize that diversity" (interview).

The New Technologies

Until recently, the development of hybrid wheat was blocked by the extremely labor-intensive task of emasculating individual wheat flowers. Since the wheat flower is perfect (it contains both male and female organs), this is no simple chore. In fact, given the relatively low value of each wheat plant (compared with tomatoes; see chapter 5), it is econo-mically impossible to produce hybrid wheat in this manner. In contrast, corn has separate male and female flowers making emasculation, or detasseling, relatively simple. Hence, hybrid corn has been available for more than 50 years.

Currently, farmers often grow and sell wheat seed to neighbors. One scientist noted that a wheat variety developed by Cargill in the late

1970s was a superior yielder in many environments. Over the next five years, acreage planted to that variety increased steadily while the volume of seed sold each year decreased to the dismay of the company (interview). In contrast, hybrids would lead to a higher volume of seed sales. Companies have attempted to justify the greatly increased prices of hybrids by arguing that they need to recapture research and development costs, and that because of the "uniqueness" of their product farmers are willing to pay the increased price, that is, the price the market will bear.

To date, the private sector emphasizes hybrids more than the public sector. The discrepancy is particularly acute for wheat. Virtually two-thirds of all research and development in wheat is done by the private sector, with the majority of the work, until recently, focused on hybridization. We identified only one public sector breeder who worked on hybrids. Seed companies wanted to develop wheat hybrids because most large wheat farmers only buy seed every three to five years, saving part of each harvest for use as seed the following season. A number of major seed companies that mounted considerable efforts in the mid-1980s to create hybrid wheats have now abandoned them because of serious technical problems.

Because hand emasculation and pollination of wheat on a commercial scale is impractical, many companies have attempted to develop genetic and/or chemical means to produce hybrids. The first step is producing male-sterile lines, either by using cytoplasmic male sterility (CMS) (based on the observation that the cytoplasm of certain female egg cells yield offspring that have sterile male parts) or chemical hybridizing agents (CHAs). Obtaining a true CMS plant is a lengthy process. Drawbacks to some CMS wheat strains include reduced germinating capacity, higher cold sensitivity (Jost et al., 1975), premature sprouting of seed while still on the plant, and lower yields. Moreover, restorer lines are even more difficult to create because the restoration of fertility is not based on a single gene but, rather, on numerous genes that modify and/or inhibit fertilization.

Another problem with CMS is the potential sterility in the F_1 progeny, which will not produce seed unless the plants have fertility restorers or receive foreign pollen. Since the farmer grows only one variety at a time, fertility must be restored in the F_1 in order to have a crop to harvest. Fertility restoring, or Rf, genes have been found in *Triticum aestivum* and related species (Hughes and Bodden, 1977). However, the genetics of male fertility and fertility restoration is complex in some varieties, involving multiple genes and/or multiple alleles for individual genes. Breeding restorer lines (R-lines) is very difficult because both genetic background and environmental conditions can affect expression of Rf genes (Sneep et al., 1979). There is also the problem of inadequate fertility restoration in the R-lines and the F_1 hybrids. Both Cargill and Hybritech Seed International (owned by Monsanto) had overcome some of these difficulties and had developed hybrid wheat varieties using CMS and R-lines, which became commercially available under the brand names of Bounty and Quantum, respectively. Recently, Cargill ceased sales of Bounty, leaving Quantum as the sole hybrid wheat on the US market.

An alternative to CMS and R-lines is the development of CHAs which, to be useful, must kill male gametes (pollen) but not female gametes. The invention of gametocides that can be sprayed on wheat and that inhibit the development of the male portion of the flower has proved a significant aid in hybridization, although the process demands spraying at just the right moment in the growth cycle of the plant. A number of chemical companies have developed gametocides and are using them to create proprietary hybrids. This is a difficult task, as one breeder explains: "Not every wheat variety responds to gametocides in the same way. First we need to find varieties that respond well to gametocides. Then we could cross them with CMS varieties. Then we would have a female population specifically adapted to being female and a male population specifically adapted to being male" (interview). Although in principle hybrids could be more easily developed from these gene pools, in practice this task could prove to be difficult.

A major problem with gametocides is that their action is often specific, that is, they sterilize florets at a certain stage of development. Since there are differences in stage of floret development within a head, a plant, and a field, a number of applications are needed. Another problem with CHAs is that their use must receive government approval, an expensive and time-consuming process. Although most published research has been done with Ethrel (Hughes et al., 1974; Dotlacil and Apltauerova, 1978), which is very specific in action, recent work has focused on other chemicals whose action is more general (Jan and Rowell, 1981).

These chemicals are often produced by large chemical companies such as Rohm and Haas, Monsanto, Sandoz, and Royal Dutch Shell, all of which have had major research and development programs in wheat. For example, Rohm and Haas developed the artificial plant hormone Hybrex, which has successfully been used to produce hybrid wheat. Several red winter wheat hybrids, both hard and soft, were produced by seedsmen and were available commercially for several years under the brand name Hybrex (Cisar, 1983; interview). Recently, Monsanto purchased the Rohm and Haas seed division and merged it with Hybritech; the sale of Hybrex was discontinued. Shell developed a CHA that produced wheat hybrids in the laboratory. Shell field tested the material in 1983 and planned, although unsuccessfully, to use it on a large scale in 1984 or 1985 (Alper, 1983).

In addition to CMS and CHAs, hybrids can also be produced via seed apomixis, in which embryos form directly from diploid maternal tissue, with the seed producing a clone of the parent. Hybrid production via seed apomixis has advantages over other methods. In principle, apomictic hybrids are cheaper to develop from a research standpoint and would take substantially less time (perhaps as short as a year) to bring to market. Since seed apomicts reproduce asexually they do not exhibit hybrid breakdown; the farmer could thus save seed to plant the following year. Finally, seed apomixis is known to occur in a number of basic grains — such as corn, wheat, rice, sorghum, and millet — as well as in citrus fruits (Mooney, 1980) and bluegrass. However, creating wheat apomicts is a difficult problem of basic biology because multiple genes are involved. Although hybrid production via seed apomixis may be biologically desirable, it is economically undesirable because the farmer would not need to buy seed every year. As a result, neither the public nor the private sector has shown much interest in apomixis.

Wheat hybrid yields first surpassed those of varieties in university trials in 1982. Since then, hybrid yields have been at least five bushels per acre higher than commonly grown varieties (Johnston, 1985), although the best performance appears to be at locations where the yields for both experimental and control plots are highest. In addition, the early hybrid work was based largely on small-plot trials where the faster-growing hybrids could draw on the water and/or fertilizers in neighboring plots containing standard varieties. In large fields these differences disappear. Moreover, in most countries, including the United States, wheat seed is produced by farmer-dealers under contract to seed companies. These farmers are unlikely to continue hybrid seed production if they do not show a profit. Three hybrids were marketed until recently: Bounty (Cargill), Hybrex (Rohm and Haas), and Quantum (Hybritech). According to some claims, hybrid wheat yields are 25–30 percent higher than yields of currently-used varieties (Grall, 1986; cf., Tudge, 1988), although this is probably an overestimate of the differences achieved in farmers' fields.

It is unlikely that the public sector will remain in the forefront of wheat breeding if significant improvements are made through hybridization. As one private sector administrator put it: "A bigger question than gene transfer is heterosis and how do you go after it. The problem is that the public sector doesn't have the scale to do this" (interview). Moreover, there are at least some who think that the public sector should abandon wheat breeding entirely. One experiment station breeder stated: "It is getting more and more difficult to get public varieties on markets. The private companies promote their varieties. There is some feeling that the private sector should develop [all] varieties" (interview).

A long-term goal behind hybridization is the development of what is called functional attribute wheat (Moshy, 1986). Even in the 1920s, the differences between high- and low-protein flour began to become more distinct. This was, in part, due to the rising living standards and the consequent demand for more cakes, pastries, and cookies (Alsberg and Griffing, 1928). The current effort is to extend further the already extant varietal differentiation found in wheat through the use of the new biotechnologies. The focus of this research is not on agronomic problems, but on end uses for the wheat. For example, wheats might be developed for easier milling or faster rising and baking. Special flavor or starch characteristics might also be developed. Farmers might be willing to pay a substantial premium for those wheats because the crop could also be sold at a premium. All of this will require considerable advances in genetic recombinant techniques with wheat. To date, transfer vectors developed for other plants have not worked for wheat (or most other cereals). However, research is now underway to develop alternative gene transfer techniques (Gasser and Fraley, 1989). Which of these will prove both scientifically feasible and economical remains to be seen.

Recently, economical methods have been developed to produce wheaten starch, a product of the biotechnological process in which gluten and starch are separated. Wheaten starch has already had a noticeable impact on the potato and maize starch industries in Europe (Bijman et al., 1986). Since the European Economic Community (EEC) countries tend to produce wheat with low gluten content, the gluten that is a byproduct of the wheaten-starch making process can be added to existing low gluten flour to make a satisfactory bread. As late as the 1950s, Britain imported 80 percent of its wheat flour from North America for this reason; today, because of breeding work and added gluten, Britain imports only 20 percent and it may soon import none (Tudge, 1988).

The advent of biotechnology has made possible the sequencing of the entire wheat genome. If currently planned research (Anderson, 1987) is successful, the entire genome will be sequenced and this will speed up the release of improved varieties. Moreover, it will undoubtedly increase the pressure on the already taxed germplasm banks to provide a wide variety of materials with detailed documentation.

To date, the major bottleneck in applying the new biotechnologies to wheat remains the difficulty in growing plants from tissue or individual cells. While this has been done for some cultivars (Scowcroft et al., 1984), it remains less commonplace than with tomatoes (see below, chapter 5). Moreover, the complexity of the wheat genome and the relatively small number of genes identified make progress slow. As Austin et al. (1986:61) stated: "At present, therefore, we do not have sufficient knowledge of the physiology and genetics of growth and development to identify 'target' genes for modification and/or transfer."

As the new biotechnologies are used more frequently and more effectively with crops as ubiquitous as wheat, one area of potential concern is the accidental creation of toxic or carcinogenic varieties. Even now, a small percentage of the human population is adversely affected by ingesting wheat products. One food scientist recently said, specifically in reference to wheat, "The protein molecule with its variety of reactive amino acid side groups is such a potential site of unexpected reactions with added chemicals that we cannot be too careful in checking that none of these units is converted into a molecule that could have some subtle but ghastly effect" (Carpenter, 1975:94).

Conclusions: The New Politics of Wheat

A number of themes emerge out of our brief review of the growth and development of the wheat industry and the concomitant development of wheat breeding. First, it should be fully apparent that nature is not a constant limiting wheat production. While one cannot grow wheat at either the North Pole or in the humid tropics, it should also be apparent that climate is a variable in determining wheat production. The same may be said of soil conditions and the character of the wheat plant. It would not be an exaggeration to say that the present-day wheat plant has been "socialized". It is as much a product of human nurture – Darwin's contemporaries called it domestic selection – as it is of nature.

A second theme of significance is the key role that the state played worldwide in the formation of wheat markets. Initially, the state provided free land, railroad and highway infrastructures (either directly or through subsidies), an inspection and grading system, a network of storage elevators in some nations, and a system of agricultural experiment stations where new varieties could be selected or bred. More recently, the state has provided credit to various parties in the production and processing system, farm programs to ensure adequate (and sometimes much more than adequate) income to producers, and guaranteed markets for surpluses unsalable internationally. One observer, nearly twenty years ago, suggested that "much of the wheat economy has acquired many of the attributes of a public utility" (Goldberg, 1968:26).

The state, of course, was not the only important player. The breeders, seed companies, farmers, railroads, elevator companies, millers, bakers, wholesalers, and retailers have all participated in the creation, modification, restructure, and collapse of various wheat markets. However, not all of the actors have had equal roles. In general, farmers have had the most difficulty in making their desires felt. As actors in a monopsonistic market, they have rarely acted as a group, and the largest have generally

Table 4.1   US white pan bread marketing spreads, 1919–88 (cents per pound)

Year[a]	Retail price	Retail spread	Baker spread	Miller spread	Other spreads[b]	Farm value	
						All ingredients	Wheat
1919	10.2					3.4	2.8
1920	11.7					3.4	2.9
1925	9.4					2.3	2.0
1930	8.7					1.3	1.0
1935	8.5					1.4	1.1
1940	8.0					1.1	0.9
1945	8.8					2.2	1.8
1950	14.3	2.6	7.0	0.6	1.1	3.0	2.4
1955	17.4	2.6	9.4	0.7	1.5	3.2	2.4
1960	19.8	3.8	10.9	0.8	1.5	2.8	2.3
1965	20.8	4.2	11.2	0.6	1.6	3.2	2.6
1970	24.2	5.4	12.8	0.5	1.9	3.4	2.6
1975	36.0	4.5	21.3	0.7	2.7	6.8	4.5
1980	50.9					5.3	4.5
1984	54.2	6.8	37.7	1.2	3.4	5.0	4.4
1985	55.3					4.8	4.1
1988	61.3					4.8	4.1

[a] No data were collected prior to 1919. Spreads by sector were not collected prior to 1947 or after 1984. Figures for 1984 are for the first quarter only. After 1981, slight changes were made in the calculation of price spreads (see Schnake, 1982).
[b] Other spreads include transportation, handling, merchandising, processing nonfarm ingredients, and cost to baker of nonfarm ingredients.
Sources: USDA, 1957, 1974, 1976, 1984a, 1989

benefited more than the smallest from the programs enacted by the state.

In addition, farmers, unlike other players in the market, have had little or no ability to increase the value added by wheat production. As Goodman, Sorj, and Wilkinson (1987) recently suggested for agriculture in general, the "wheat sector" has been largely transformed by industrial capital through both appropriation (i.e., the transformation of certain agricultural activities into industrial ones) and substitution (i.e., the removal of activities previously accomplished on the farm to industrial settings). This, in turn, has moved much of the capital accumulation in agriculture off the farm. Consider, for example, the "value added" (i.e., the value that is added to wheat) in the process of bread making, as depicted in table 4.1. Several trends are apparent. First, the nominal farm price of wheat has risen, although more slowly than the price of a loaf of white bread. Second, the baking and retailing processes account for most of the change that has occurred in value added over the years. This should not come as a surprise as these actors have the capability of increasing value added through improved quality (real or perceived), brand name identification, introducing (monetary) value-enhancing changes in the loaf itself, and improving packaging, thereby increasing shelf life and sales. In addition, though not reflected in the figures presented in the table, there has been a marked tendency toward product differentiation and the creation of new wheat products. These avenues are not generally open to the farmer, irrespective of farm size.

This is not to suggest that the nonfarm sector of the wheat industry has been particularly profitable. It is simply that these industries have a larger and growing spread between the price of the raw materials they buy and the price of the product they sell. This is generally accomplished by investing more capital in the processing or, in rare cases, more labor. Given the peculiarities of food processing and marketing, it does not necessarily mean an increased profit. In fact, the food processing, wholesaling, and retailing industries have been shown to operate at singularly low margins, although their total profits are high as a result of their enormous size. The profitability of grain export firms is more difficult to observe as the largest companies are all privately held multinationals. As early as 1960–7, six large firms handled 90 percent of US wheat exports (Goldberg, 1968). Over the years, the tendency has been for these companies to integrate vertically into both elevator operation and flour milling. Some see them as particularly profitable institutions (e.g., Morgan, 1979). However, so little is known about them that it is of little value to speculate here.

Among the many issues negotiated by these diverse actors, the nature of wheat was and remains of primary importance. What is wheat to be? Is it red or white? Hard or soft? Freemilling or not? These and related questions have been the province of the breeder. Yet, it should be apparent that the breeder has rarely been free to decide the answers to such questions alone. Such decisions have been made in a highly politicized environment (although one usually not recognized as such by the breeder), in which various interests attempt to restructure the wheat plant and thereby the social world through alliances and bargaining. The breeder who ignores the power relationships among these interests may produce a new variety which, although it suits one or more interests, will not be used. This is not to say that the breeders have no control over the process at all. Through their participation in the negotiation and persuasion process, their voices can be heard too. As the only actors intimately familiar with the range of possible breeding outcomes and tradeoffs, breeders can restructure the various interests, and encourage the formation of new alliances and coalitions and the dissolution of others. Such "political" work goes hand in hand with the technical work of breeding itself. It appears that breeders are only reorganizing the genetic structure of the wheat plant, yet it is both the wheat and the other actors in the system that are reorganized.

In the last several decades, however, the picture has begun to change. In the past, farmers, millers, elevator operators, and bakers have all been, with only a few minor exceptions, distinct actors in the wheat market. Now, it appears that a new set of changes, made possible by the patenting of plant materials and the development of plant molecular biology, are beginning to reorganize it in ways never before possible. Millers and elevator operators are now sometimes linked, grain companies now own seed companies, retailers own bakeries, and breeders are employed by grain companies. While the precise details of this restructuring are difficult to predict, the general outlines are fairly clear:

1 Farmers as actors in the system of wheat production will continue to lose their influence. This will occur since farmers appear unable to develop their own vertically integrated institutional structures, despite the long existence of farmer cooperatives. Moreover, the monopsonistic character of the markets for both wheat and farm inputs will continue to put farmers at a disadvantage. While some improvements that benefit farmers will undoubtedly appear, it is likely to continue to be the larger producers who are the major beneficiaries. In short, we must agree with Buttel (1981:18) who writes: "There are unlikely to be appropriate technologies developed which will enable a 100-acre wheat farmer in the Great Plains to outcompete his neighbor with 5,000 acres." In addition, although farmers are numerous by comparison with other actors in the wheat market, they are of trivial importance in national politics. Hence, while appeals to agrarian values are likely to remain a part of the public relations efforts of large food companies and farm legislators, the legislation is likely to be of less and less value to wheat producers. Finally, overproduction is likely to continue for the foreseeable future.

2 Greater vertical integration will make possible greater control over the wheat market by a small number of national and multinational entities. Since the profit margins appear rather small in individual segments of the wheat industry, these vertically integrated firms are likely to focus on the development of new products and processes that can be protected either as trade secrets or by the new patent laws. Although it is not likely to be developed soon, functional attribute wheat, bred with specific technical needs in mind, is clearly of tremendous value to these firms. In the meantime, these firms are likely to exert greater influence over the public-breeding programs by encouraging more "basic" (molecular) research, while private seed companies continue to do more of the conventional breeding and development of hybrids.

3 The role of the state is unlikely to diminish in the near future. The state may create costs that otherwise would not be incurred. However,

the state ensures plentiful supplies, subsidizes storage and export, enforces standards, and pays for the basic research. While some of the actors in the system may wish to bend or even break some of these rules and laws, integration is likely to increase support for state intervention for the stability that it ensures. Farmers may not be happy with this, but as noted above, their collective power is diminishing.

4   As the industry becomes more integrated – vertically and geographically – it will be more and more difficult for those outside the industry to comprehend it. Barring a major public outcry, it is unlikely that a study of this type will be feasible in 20 years because the necessary data will simply be too hard to obtain.

5   Finally, barring major legislative initiatives, consumers are likely to remain the passive recipients of changes in bread and other wheat products. It strains the imagination to believe that American tastes in bread were as uniform and ubiquitous as a loaf of white pan bread until just a few years ago. It is more likely that such bread was considerably cheaper than bread from local bakeries and was seen as an adequate substitute for more expensive wheat products. In contrast, the French ensured the existence of great variety by putting a price floor on bread. The American desire for more variety and taste has led to great product differentiation in recent years. This, in turn, is undoubtedly related to rising income levels.

5

Tomatoes:
The Making of a World Crop

Tomatoes offer an excellent contrast to wheat. Unlike wheat, tomatoes have been and, in most places, remain a garden crop. Although a perennial, they are cultivated everywhere as an annual. Like wheat, tomatoes rarely cross-pollinate; only 0.5–4.0 percent do so in temperate zones (Tigchelaar, 1986). In the United States, and probably in several other Western nations as well, they now constitute the major vegetable source of essential nutrients in the diet (Nevins, 1987). Worldwide, tomatoes are the most widely grown vegetable after potatoes. Over 45 million metric tons are marketed annually (Villareal, 1980). Tomatoes are eaten fresh as well as in processed form. As a result of the large-size fruit and the higher ratio of seeds produced by each plant, they have been the subject of much hybridization research. Tomatoes are also a model crop for plant genetic studies and have been the subject of much scientific analysis since the entry of molecular genetics into the plant sciences during the last decade. Moreover, tomatoes are a New World crop that crossed the Atlantic going east to become one of the world's most popular vegetables. We begin by examining their history.

History

Tomatoes apparently originated somewhere in Central America and were used as food by the indigenous peoples of that area. There is some disagreement as to the exact point of their origin, with Mexico, Guatemala, and Peru mentioned by various authors (Villareal, 1980; Tigchelaar, 1986; Nevins, 1987). These early landraces, however, were small, with fruit no larger than a modern cherry tomato. With the Spanish conquests of the sixteenth century, the tomato was moved from the New World to the Old. As early as 1523, Cortez brought seeds to Europe. It is these seeds that are the origin of all varieties currently grown in Europe and Asia (Villareal, 1980). English authors described the tomato as an ornamental as early as 1578 (Morrison, 1938), gracing the gardens of the wealthy. In France and Italy, too, the tomato was grown as an ornamental. It soon became known (for reasons that are unclear) as the "love apple". However, being a member of the nightshade family in which many poisonous plants are found, the tomato was widely regarded as poisonous. Therefore, tomatoes did not begin to appear in European vegetable gardens until the early nineteenth century. They were not widely recognized in Europe until 1822, although Boswell (1937) argues that they were grown on a field scale in Italy by 1800.

Ironically, despite their New World origins, tomato production in the United States had to await their reintroduction from Europe. The first documented introduction into the United States was made in 1710 (Tigchelaar, 1986). They remained rare enough for the next three-quarters of a century that Thomas Jefferson planted them as a novelty in his garden in 1781. Despite this, more than 50 additional years passed before the tomato became widely known in North America.

Once known, however, tomatoes rapidly became a highly prized vegetable. By 1863 some 23 cultivars were known in the US. The first American variety was Tilden, developed by Henry Tilden of Davenport, Iowa, in 1865. By 1905, L. C. Corbett, a plant scientist at USDA, could rightly remark that, "the tomato is one of the few garden vegetables of American origin holding high rank as a commercial crop which has come into general cultivation within the last century" (Corbett, 1905:5). The high regard for tomatoes is evidenced by the introduction of an early variety named Trophy in 1870 by a Dr Hand of Baltimore (Boswell, 1937). Early maturity was highly valued by northerners because their short growing season limited production. Trophy sold for the astounding price of $5.00 for a packet of 20 seeds (Morrison, 1938)!

In Europe, too, by the end of the nineteenth century, tomatoes were widely used. A US consul in Europe said at the time that "tomatoes were but a short time ago an article of luxury in Great Britain, only used for the pampered palates of the rich; but now they have become a common dish on the table of the working classes" (Hanauer, 1900:394). He went on to note that, in Germany, canned tomato pulp was available cheaply, while fresh tomatoes (shipped in from southern France where they were apparently plentiful) were only to be found in the best hotels.

While only a few varieties were available in the early nineteenth century, by the latter half many seedsmen had begun to develop new ones. The most famous seedsman, A. W. Livingston, realized that the superior way to produce desirable varieties was to select not the best fruits – as his contemporaries had done – but the best plants. Between 1870 and 1893 Livingston developed 13 new cultivars, all of which were widely sold. During this period the number of varieties soared from a handful to several hundred, in part due to the importation of new stocks from Europe, the work of men like Livingston, and the tendency toward duplicate naming. In 1886 and 1887 the horticulturist Liberty Hyde Bailey, then at Michigan State College and later to become the renowned Dean of the College of Agriculture at Cornell University, felt it necessary to conduct field tests of the plethora of cultivars available (Bailey, 1886, 1887). Bailey collected 170 samples which he carefully grew out; he concluded that the group represented only 61 distinct varieties. No doubt Bailey's research was welcomed by confused farmers and gardeners, although it is unclear just how it was greeted by seedsmen. A similar study conducted in 1936 identified only 35 varieties (Morrison, 1938).

By the turn of the century, tomato production was a highly organized, complex business with separate canned and fresh markets. In 1898, Edward B. Voorhees remarked that "the tomato is grown more largely for canning than any other vegetable used for this purpose" (Voorhees, 1898:7). When destined for canning, tomatoes were often grown under contract. Initially these contracts were quite simple and straightforward. An early USDA bulletin on tomatoes recommended that growers have such contracts with canners and praised the form used by most New Jersey growers:

This is to certify that we _____ have bought of _____ the product of _____ acres of tomatoes for the season of _____ at $_____ _____ per ton, delivered at our cannery at _____ _____. Stock to be in first-class mercantile condition. To be planted about _____. (Corbett, 1905:31)

At that time 300,000 acres were in production in the United States, with Maryland the leading producer followed by New Jersey, Delaware, and Virginia. Growers would often begin the season by selling their early crop on the fresh market, and switching to the processors only after the fresh market price had fallen below contract price (Beattie, 1921). In examining the brief contract form above, what is remarkable from our present-day perspective is the lack of discussion of fertilizers, pesticides, or even seeds. Of course, at the turn of the century, the use of chemical fertilizers and pesticides was not yet widespread. To the contrary, seed varieties were well known and were identified with certain agronomic characteristics. However, the choice of variety was apparently still a problem. Voorhees (1898:18) noted: "In the matter of varieties, ... too much dependence should not be placed upon the name or upon the fact that a neighboring farmer secures good results from a given variety, since there are so many variations in the character of soils, . . . that the best variety is usually one that is, in part at least, developed by the individual grower." Only much later, with the development of more sophisticated plant breeding techniques, did this confusion over varieties gradually disappear. At that time, not only did seed varieties enter into the contracts between growers and canners, but new distinctions were gradually made between processing and fresh market tomatoes. These distinctions involved not merely the end user but the entire social organization of the industry.

While 300,000 acres of tomatoes may seem small on a national scale, it should be remembered that tomatoes were a highly labor-intensive crop to plant and harvest. Indeed, tomatoes are still recommended for production today in places where land is scarce and labor is abundant (Villareal, 1980). In the northern climates, plants were often started in greenhouses or hothouses and then set out after the threat of frost was gone, giving them a few extra weeks growth during the season and increasing yields substantially. At harvest, labor needs peaked: "The fruits should be gathered two or three times a week if the tomato is grown as a truck crop. When used for canning purposes the harvesting periods need not be quite so close" (Corbett, 1905:13). Harvesting requires an enormous quantity of cheap, unskilled labor for a relatively short time. Some labor was provided by children on summer vacation, but much of it was furnished by migrant laborers who worked the harvests starting in Florida and moving north over the season. A similar, though initially smaller, stream began in southern California and worked its way up the west coast. Bulletins issued by USDA and the universities say little or nothing about the unskilled laborers' situation.

Greenhouse production was widespread enough by the turn of the century that Voorhees felt obliged to provide a separate section on the subject in his *Farmers' Bulletin*. Yet, such production never reached the levels achieved in Europe, in part, due to the ease of shipment of off-season produce from Florida up the east coast to northern markets.

Today, there is great variation in the types of tomato products available. In addition to fresh tomatoes (which include cherry tomatoes and various more exotic types), there are numerous canned products, including whole tomatoes, chili sauce, juice, pulp, paste, catsup and soup. There are also many specialty products, such as cocktail sauce, produced only in small quantities.

Although Americans and Europeans limit their consumption largely to canned and ripe, fresh products, many consumers in other nations have developed alternative ways to eat this rather adaptable vegetable. The Taiwanese, for example, often prepare tomatoes as a candy. Other products include tomato powder and pickles. Some nations also distinguish two types of fresh tomatoes: those consumed as a fruit and those used for home cooking (Villareal, 1980).

The tomato is an enormously adaptable plant. It can satisfactorily produce in tropical climates, from which it originates, as well as in fairly cool temperate climates. Unlike wheat production, which is confined almost entirely to large fields, tomatoes may be, and are, grown anywhere from fields of thousands of acres to window boxes of urban dwellers. Also, unlike wheat, tomatoes are now grown in nearly every nation in the world. As one might expect, such widespread production means that there is great variation in the way in which tomatoes are produced, processed, and consumed. There are national differences in variety preferences as well. However, there is one sharp division of overriding significance in understanding tomatoes as a commodity: the distinction between fresh market and processing tomatoes. Hence, we discuss these two types separately, almost as if they were different crops.

Processing Tomatoes

Production

In most of the world, the production of processing tomatoes is a secondary or part-time activity of farmers. California is the exception to this rule, where nearly all processing tomatoes are produced on large farms that specialize in this production practice. As noted above, this is mainly a consequence of the mechanization of tomato production, a change in its infancy in Europe and uncontemplated in most of the rest of the world. Nearly everywhere, however, tomatoes for processing are grown under contract with a processor. Today, the contract specifies what seed is to be planted and the date the product will be delivered to the factory. It may also specify that certain agronomic practices be used. From the processors' point of view these terms are essential as they wish to spread the processing time out as much as possible so as to get the maximum use out of their equipment. In this section we first examine the rapid growth of large-scale processing tomato production in California and in the US in general, and follow this with a brief comparison with three European producing nations.

In the US, processing tomato production has traditionally been divided between several eastern and midwestern states and California. With the opening of the vast irrigation systems in the California valleys during the first decades of this century, and the improvement in the

speed of rail transport, the tomato industry began to move west. The
California valleys offered virtually the ideal climate for tomato production once the missing factor, water, was added. By the 1940s, California
was the leading producer of both fresh and processing tomatoes (Porte,
1952). In 1950, 36 percent of the US tomato tonnage was grown in
California; by 1975 that figure had jumped to 85 percent, with many
eastern tomato farmers leaving farming or turning to other crops. This
occurred at the same time that tomato consumption per capita in the
United States rose 68 percent, between 1948 and 1975, peaking in 1980
(Stevens, 1986).

The industry shifted to California for other reasons as well. Perhaps
of greatest importance was the mechanization of the harvest which
began in the mid-1960s. As the central theme of this volume is biological rather than mechanical technologies, we shall not dwell on the
various details of the construction of the tomato harvester. Instead,
we will only discuss the tomato harvester as it changed the system of
tomato breeding and production (and was changed in turn by it).
However, the reader should be aware that tomato mechanization has
been paradigmatic of disputes over the proper role of publicly supported research in the US for some time now (see Schmitz and Seckler,
1970; Hightower, 1973; Friedland and Barton, 1975).

Although the mechanization process for harvesting tomatoes began
more than 100 years later than it did for wheat, the fundamental
reasons for it were similar. The high cost and difficulty of management
of large field labor forces was clearly a great incentive. By the late
1940s, agricultural engineers at the University of California at Davis, as
well as at several midwestern universities, had begun to develop prototype harvesters. The program, however, was hindered by the fragility
of the tomato fruit. Only with the development of a new machine-
harvestable variety would a major shift become possible.

The immediate cause for adoption of mechanical harvesters was the
termination of the Bracero program in 1965. The Bracero program had
been authorized by a 1951 amendment to the Agricultural Act of 1949.
That amendment (12 July 1951, ch. 223; 65 Stat, 119) permitted large
numbers of Mexican workers to be recruited to work in California for
the fruit and vegetable harvests there, returning to Mexico (at least in
principle) after the season ended. The termination of the program
meant the end of cheap foreign labor, and it was a strong impetus to
mechanization. Adoption of the harvester reduced the labor time
needed to harvest an acre of tomatoes from 113 hours to only 61, a
reduction of nearly 50 percent (Gould, 1983). By 1970, just five years
after the Bracero program's termination, 100 percent of the growers
had shifted to mechanical harvesting. This occurred despite the fact that
the initial mechanical harvesters left 15–20 percent of the fruit in the

field.[1] This rapid adoption is all the more startling when one considers the sizable expense incurred for highly specialized equipment used only a few weeks a year, and the fact that the vast majority of tomato farms were too small to utilize the harvester profitably.

The harvester design that was finally adopted consists of four parts: a pickup mechanism, a fruit and vine separator, a hand-sorting area, and a discharge or container loader (Gould, 1983). In operation, the mechanical harvester pulls the entire tomato vine from the ground and runs it through the machine, removing the fruit which is then hand sorted by workers on the machine. The workers remove the green tomatoes, damaged fruit, and trash that the machine misses as the fruit travels on a conveyor belt. Thus, the machine is properly seen as a (dis)assembly line that has been made mobile to compensate for the necessary spatial spread and fixity of the growing plants.

Use of the harvester has reduced the need for hand labor in the fields and has speeded up the harvest itself. However, it accomplished this only by restructuring the industry and creating a new set of problems that required additional scientific and technical expertise to solve. For example, machine harvesting substantially increases the quantity of soil that is collected with the fruit. Not only must the soil be removed, but it may contain bacteria that can cause fruit spoilage. Fruit damage is more likely, too.

Moreover, much as the adoption of the three-field system in the Middle Ages required restructuring the landscape, replacing square plots with long rectangles, the mechanical harvester also requires that farmers rearrange their fields. Gould (1983) recommends that rows be not less than 600 feet in length in order to minimize turning the large harvesting equipment. In addition, the equipment works best on flat, well-graded fields; hilly land is not machine harvestable. Moreover, since the equipment is of no value for any other crop, it promotes crop specialization. This can be seen in the rapid increase in the average size of California processing tomato farms. Whereas the average tomato farm had 91.1 acres in 1956, by 1975 it had 361.7 acres (Brandt et al., 1978). At the same time, the number of tomato growers declined from about 4,000, before the widespread adoption of the mechanical harvester, to less than 700 in 1972 (Friedland and Barton, 1975). Moreover, a complex set of events, including the development of the harvester (which compensated for a lack of local labor), the creation of new irrigation canals, and the construction of a new superhighway, moved the center of tomato production to the southern end of the Central Valley.

---

[1] Rasmussen (1968) notes that while growers were initially concerned about high loss rates, Lorenzen argued that hand harvesting was no more effective (see p. 158).

Mechanical harvesting, however, is not the end but the beginning of the story of California's dominance in the tomato industry. Mechanization provided an immediate advantage over the hand harvesting of the eastern and midwestern farmers as the wet summers there precluded or slowed mechanization in most places. The heavy machines would bog down in the mud. Hence, farmers were reluctant to mechanize. Indeed, while mechanization was embraced by California producers, it was resisted by midwestern producers. Only in the face of a farmworker-organized boycott of the processors did harvest mechanization begin to occur. In an attempt to stop the boycott, processors gave preferential treatment to farmers who mechanized their harvests (Rosset and Vandermeer, 1986). Even today, only 70 percent of midwestern tomato farms are mechanized; Vandermeer (1986) suggests that hand harvesting is still more desirable there, given climatic conditions and the higher loss rate (i.e., lower yield per acre) of mechanical harvesting.

Mechanization also conferred other significant advantages on producers. The lugs (small containers holding about 50 pounds used to transport tomatoes from the fields) were soon replaced by bins that held 800 to 1,000 pounds of tomatoes each. Bins, in turn, have been replaced by bulk-type gondolas, introduced in 1970, that hold 25–30 tons of tomatoes each. These huge containers have their own problems. For example, the gondolas must be filled more carefully to avoid damage to the fruit. In addition, the tomatoes must be properly distributed in the trailers so as not to affect the balance of the trucks during hauling. Finally, immediate shipment to the processing plant is essential as the tomatoes heat rapidly under such conditions (Friedland and Barton, 1975). Nevertheless, by 1976, 65 percent of the tomatoes were delivered in gondolas to the processors.

Still another technical change associated with harvest mechanization is the electronic or optical sorter. Introduced in 1976, the electronic sorter has the capability of distinguishing between green and ripe tomatoes by light diffraction. Thus, the sorter is a good substitute for human sorters as long as the problem is limited to color. However, the sorter cannot remove defective, mechanically injured, or moldy fruit. The sorter is most effectively used at the beginning of the season when the major problem is green tomatoes. Even with the electronic sorter, however, some hand labor is required to compensate for the machine's inefficiencies.

The harvester has also restructured the labor force used for harvesting, since considerably fewer persons are now needed for the harvest. Male Mexican farmworkers have been replaced by women from local communities who stand on the harvesting machines (Friedland and Barton, 1975). Since these women are from local communities, it is possible that employment may have increased locally at the expense of the Mexicans. In any case, without any data whatsoever, Gould (1983:69) assures us that "it has been proved that women are more efficient than men for sorting."

Brandt et al. (1978:95), in a publication of the University of California, have argued that "our calculations suggest that added employment due to expanded mechanization-induced output was about as great as the loss of harvest labor jobs. Since the workers filling these jobs came from different populations, the welfare implications of this shift are not clear." In contrast, Friedland and Barton (1975) argue that there was a net loss. Even if the former group were right, their statement is rather peculiar. It is clear that the Mexican workers who returned to Mexico are likely to have remained unemployed or to have been employed at lower wages; after all, they originally came to the US because of low wages or lack of jobs in Mexico. Similarly, other states undoubtedly lost jobs as the industry became more concentrated in California. Thus, the welfare implications *are* clear. In fact, nearly a decade earlier, Thor and Mamer (1969:69) argued that mechanical harvesting "may ... have a high cost in human suffering if we do not recognize that it could greatly reduce employment opportunities for the unskilled workers".

The hand-sorting process does require coordination between the driver of the vehicle and the sorters. This problem is reduced when electronic sorters are used. Also, fruit sorters must be well trained: they must not oversort by throwing away too many acceptable fruits or undersort by leaving in too many bad fruits. The former will reduce the grower's profits because of the discarding of fruits that the processor would purchase. The latter may result in rejection of the load for exceeding the limits specified in the contract or by state inspectors. In practice, growers have found that by varying the speed of the machine they can assure that the maximum permitted percentage of culls, greens, and dirt remain in the load. Undoubtedly, this must make the work of sorting rather frustrating.

In short, the mechanical harvester and associated technical changes served to restructure processing tomato production. By using the harvester, the industry became more concentrated in central California and declined in the East and Midwest. Farm size grew considerably as smaller producers dropped out, and most of the old jobs for farmworkers were eliminated while new types of jobs were created. Finally, American tomato producers were able to compete successfully with those in other nations using much cheaper labor supplies.

Machine harvesting also added to the complexity of the growing process because it created stringent demands for uniformity in the fields to conform to the needs of the harvesting equipment. Some of this uniformity is bred into the plant itself but much depended on the growers' preparations. Friedland and Barton (1975:11–12) explain:

The plants should be distributed uniformly along the rows if the machines are to harvest the maximum yield of a field. The rows and beds must be uniform if the machine is to move through the field without damaging the plants. The seeds must be planted at proper depths or there will be sporadic emergence of plants ...
Tomato vines must be ready for harvest simultaneously. A large grower with several thousand acres of tomatoes does not want all fields to ripen at once; once ripe, tomatoes must be harvested quickly if quality is to be ensured and waste and spoilage are to be avoided. Within a given field, however, all fruit should reach maturity over a short period.

In California, with its nearly ideal growing conditions, tomatoes are seeded directly. In other parts of the US and in most of the rest of the world, small seedlings are grown in trays and these are then transplanted to the field. (In fact, in many places, numerous nursery businesses exist to supply seedlings to tomato growers. In the US and other developed nations, these are largely automated.) In this sense the system used in California is actually less complex than that used elsewhere. However, other requirements of mechanical harvesting dictate a more complex growing environment. Weed and insect control, through the use of herbicides and insectides, is essential, although some manual labor is still necessary. Moreover, many of the weeds are biologically similar to tomatoes, making chemical weed control difficult.

Despite careful planning and scheduling, climatic conditions can speed up or retard the harvest. Therefore, there is always a considerable, irreducible element of risk in tomato growing. The risk increases at harvest time, when, even in regions where machines are used, there is a considerable increase in the demand for labor. In the case of those areas of the world where hand harvesting is still used, tomatoes are two to three times more labor intensive than rice, which itself is already a highly labor-intensive crop (Villareal, 1980).

After harvesting, tomatoes are graded, a process of great importance to processors and growers alike because it determines the price to be paid for the load. Given the greater variability of tomatoes, compared with wheat, the grading process is often a point of contention. Questions about the sampling procedures employed are not uncommon as these procedures are more problematic than they are for wheat. The actual grading is done by graders appointed by the state and the exact procedures vary from place to place.[2] Grading was first introduced in

---

[2] Friedland and Barton (1975:17) provide a detailed account of the grading process in California. However, these grading procedures differ from state to state and from nation to nation.

the US in 1926. In 1933, only 15 percent of the harvest was graded and 8–9 percent of the tomatoes were culled. Today, 90 percent of US tomatoes are graded and only 2 percent are culled (Gould, 1983). In addition, the USDA grading standards have been adopted by many other nations as well (Villareal, 1980).

Because of the sharp contrast in organization, it is useful to compare the highly mechanized, processing tomato production system used in California with other less mechanized systems. Such sharp differences do not yet exist for fresh market tomatoes. Examination of the situation in several other nations will also give us the opportunity to develop a secondary theme: the internationalization and growing competition in the processing tomato market. Fortunately, Ronald Y. Uyeshiro (1977) of the USDA has studied processing tomato production and processing in Portugal, Spain, and Greece, three important relatively recent entrants into the international market. Our comments on these three nations draw heavily on his work.

Spain   Spain experienced a 66 percent increase in processing tomato production in the short period from 1970 to 1975. Most of this growth was due to the international market for tomato paste. Spain has three major tomato-producing regions, two of which are particularly important for processing tomatoes. As in California, most of the production is accomplished in irrigated fields. Overall yields, however, are low (27 tons/hectare compared to US yields of approximately 65 tons/hectare) as some areas are still rainfed. Farm size in the Extremadura area averages 4–5 hectares and is even smaller in the other two regions.

Most tomatoes are planted as seedlings. Uyeshiro (1977:25) explains why: "One of the major limiting factors affecting the implementation of this technology [direct seeding] is the lack of a suitable variety to cope with wide temperature fluctuations during the early spring, which has caused problems with uniform germination." Harvesting is still done by hand on most individual holdings. Given the small size, this is perhaps not surprising. On the other hand, processors do use mechanical harvesters on their own farms. Since processors produce fully 40 percent of the crop themselves, this is of considerable importance. (Contract growers produce about 50 percent with the remaining 10 percent purchased on the open market as needed.) Labor is in short supply during the harvest season because numerous other crops are harvested at the same time. Lug boxes are disappearing and being replaced by bulk handling systems, but this has been limited by the lack of a suitable variety that will resist the rougher handling. The price of tomatoes is negotiated between the growers and the processors. However, since the processors have a better knowledge of supply and demand, they have a noticeable edge in contract negotiations. There is no grading system in Spain; inspectors simply examine the raw product and base prices on the percent deemed usable.

Portugal   Portugal presents a somewhat different story. Marglobe and Rutgers cultivars were introduced just after World War II, but neither was successful because the necessary cultural practices were not accepted. In 1957, a local manufacturer convinced Heinz to help modernize the tomato industry. As a result, the area planted soared from 1,000 to 26,000 hectares (de Oliveira, 1980). Portugese processing tomato production rose rapidly during the 1960s, peaked in 1973, and leveled off through the late 1970s, in part, due to the lack of a satisfactory export market and, in part, due to a shortage of irrigation water and rising labor costs. Most tomatoes in Portugal are grown in the flat central river valleys under irrigation. Yields have averaged 34 tons/hectare, with the better growers easily obtaining double that. The average farm size is 3–5 hectares but this masks what in reality are three distinct groups: family farms of 1–3 hectares (most of which are almost undoubtedly part-time operations), commercial producers with 4–100 hectares, and a very few processor-owned farms of 101–5,000 hectares.

As in Spain, most plantings involve the transplanting of seedlings. Also, as in Spain, there are annual negotiations between growers and processors over the price to be received. However, less than 50 percent of the harvest is either under contract or on processor-owned farms. Moreover, a grading system does exist, consisting of two prices depending on quality. As of the mid-1970s, there had been little harvest mechanization and only 10 harvesters were in operation in the entire nation. However, because of perceived labor shortages and high wages, growers were becoming interested in mechanization. Moreover, as the American-built machines were judged too large, work was underway to develop a Portuguese machine that would be about half the size of an American one. The government, however, has discouraged mechanization because of high unemployment and has even insisted that growers (and processors) use more labor. More recently, low prices have begun to force mechanization but it has been hampered by the lack of specialized labor and by the fact that many of the larger producers (who could most easily afford the harvesting equipment) have now left the industry (de Oliveira, 1980). With mechanization has also come the widespread adoption of Cal J, a variety specially adapted for direct seeding and mechanical harvesting (Calado et al., 1980).

Greece   The final case we examine is that of Greece. During the short period from 1970 to 1975 Greece increased its processing tomato production by 210 percent. This was undoubtedly due, in part, to grower subsidies provided during that period. During the period 1970–74 a direct subsidy was paid to the grower of $8.30 per ton. In 1975, with large surpluses, the government gave farmers $20 for each ton unharvested. In Greece, 80 percent of the tomato farms are from 0.5 to 1.5 hectares, and farm size is declining! This is in part attributable to the fact that primogeniture is not observed in Greece; instead, farms are divided among the children of the family.

Tomatoes are transplanted by hand as in Spain and Portugal and, as of a decade ago, no research on direct seeding was being performed. Also, given the small farm size, harvest mechanization has not been contemplated. Instead, harvesting is accomplished through three hand-pickings per season, often employing only family labor. The use of bulk bins is also limited. Instead, lug boxes are the major method of transporting tomatoes from the field. In addition, the varieties used in the late 1970s could not withstand rougher handling. As in Spain, most production is under contract. Annual negotiations take place under government direction, with the government having the option of adjusting the subsidy so as to achieve what is known as a "total guaranteed price". In practice, the government has been reluctant to let this total guaranteed price decline below the level of the previous year. Nor, as we shall see below, are these the only ways in which the Greek government supports the processing tomato industry. Nevertheless, significant further expansion appears unlikely because the cotton harvest occurs at the same time so a labor shortage would rapidly develop. Moreover, mechanization is unlikely with such small units.

Our discussion of processing tomato production in California and in the three European countries provides some indication of the wide variation in production practices, and shows some of the ways in which in all four cases production is converging toward a single capital-intensive technological system. As we shall see below, much the same is true for the processing sector.

Processing

Processing tomatoes on a large scale is a well-established industry in the United States. Today, tomatoes are received at the processing plant where they are sorted to eliminate unwanted material (e.g., trash, bruised fruit, greens) (see figure 5.1.), sized, and washed. Processors in some states may also grade tomatoes again at the factory to conform to contracted standards. In California (and probably in other places as well) growers complain that standards are given more exacting inter-

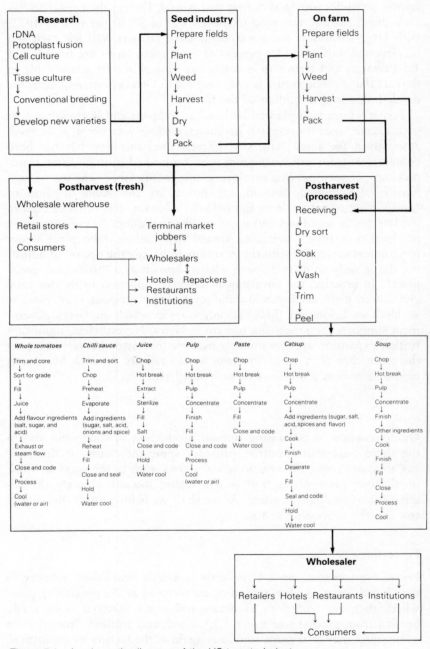

Figure 5.1 A schematic diagram of the US tomato industry.

pretations during the height of the season when the processing plant is operating at peak capacity. At that time there is more than enough produce to keep the plant operating at full capacity, so processors can afford to be somewhat more discerning than at the beginning or end of the season. However, only about 1.5 percent of all loads are rejected (Friedland and Barton, 1975).

The washing process is automated and consists of soaking the tomatoes in a large water tank followed by a spray rinse. Final sorting, trimming, and coring make up the next step in the process. Tomatoes are then treated with steam, lye, or infrared heat to loosen the skin before it is mechanically removed. At this point a final decision is made as to what use the tomatoes will be put. The major product alternatives are whole canned tomatoes, paste, puree, soup, and catsup. Whole tomatoes are packed cold and calcium salts are added to maintain firmness. Then, the tomatoes are put into containers and heated in a steam tunnel. Once they are sufficiently heated, they are sealed and cooled, usually by water emersion. The cooling creates a vacuum in the can that is sufficient to retard spoilage.

Since tomatoes are usually sufficiently acidic (pH = 4.6), it is not necessary to add anything to further retard spoilage. However, many recently developed tomato varieties are not acidic enough and require the addition of citric acid to ensure against spoilage. Otherwise, little is done to preserve whole canned tomatoes. In contrast, catsup, paste, puree, and soup are usually cooked in 250-gallon tanks to reduce water content.

The decision as to what products to produce is partly a function of expected market demand. In reality two different decisions must often be made. First, the processor must decide what to process (and consequently what area to contract for) because many plants are capable of producing both tomato products and other fruit or vegetable products. Clearly, if market demand is such that the profits are higher for producing some other vegetable or fruit that matures at the same time, then tomato acreage contracted for will be reduced. In addition, a second decision must be made as to what tomato products to produce. This decision is partly a function of market demand, but it is also dependent on the quality of the raw materials received by the processing plant. Brandt et al. (1978:27–28) explain:

> Multiproduct, multiplant operations have special advantages. In the case of multiple tomato product forms, greater flexibility in packing is obtained. Because tomato product forms require different characteristics (consistency, soluble solids, peelability), plant managers are able to alter product form allocations depending on the quality of the raw tomatoes received. Processing nontomato

products allows firms to utilize the same equipment and facilities for longer periods since the harvest seasons for these commodities usually differ.

One may rightly make an analogy here with the way that millers divide their production into flour and mill offal depending on the market for the various grades of flour. However, millers are limited to the production of wheat flour (and occasionally maize or barley) and cannot pursue the product differentiation that tomato processors can. On the other hand, the tomatoes are highly perishable whereas wheat may be stored for a considerable period of time in hopes of better market conditions. It can be argued that tomato processors have this option as well since their canned products can be stored. Nevertheless, this ignores the substantial capital that must be tied up under such circumstances. However, new technologies have been developed that clearly give the tomato processor significant advantages in comparison with the flour miller.

Many processors preprocess tomatoes into 65-gallon drums for later processing into various tomato products depending on the market. More recently, some processors have constructed aseptic tank storage facilities where semiprocessed products can be stored.[3] These practices provide considerable bonuses for the processors. First, they allow spread of processing over a longer time, permitting more efficient use of equipment. Second, they allow a more rapid response to changes in market demand for a particular product. If the price of puree is high, they can use the concentrate to produce puree; however, if there is more profit in paste, they can use it for paste. Third, they permit a more even distribution of labor over the season. Fourth, they cut down on raw product loss since the major causes of loss are eliminated. Finally, they reduce transportation costs. Today, California processors often send 20,000-gallon rail cars of concentrate to eastern and midwestern markets.

In short, drums and the new aseptic tank technology offer tomato processors a versatility that in some ways surpasses that of the millers. They permit virtually immediate response to market changes and turn a considerable portion of tomato processing into a continuous process technology. This is in keeping with the tendency of contemporary agriculture in general to move into the industrial arena (Goodman et al., 1987).

Processors are particularly concerned with what they define as quality

[3] Ironically, aseptic storage was itself encouraged by mechanical harvesting. By shortening the period of harvest and increasing the proportion of dirty and bruised fruit, mechanical harvesting created a serious bottleneck at the processing plant. Aseptic storage reduces losses from delays in processing (Vandermeer, 1982).

considerations. For processors, quality includes to some extent taste and texture, but most importantly, color. Gould (1983:229) explains: "In the case of tomatoes and tomato products, the degree of color quality practically represents the measure of total quality." He goes on to note that "the consumer associates certain color characteristics with fresh and wholesome products. Thus, the extent to which original natural color is present in the processed product is an important criterion of a quality tomato item" (1983:230). Other aspects of quality that are important to processors, although generally not to consumers, include high solids content, low pH, firmness, easy peeling, and crack resistance.

High solids content is particularly important to processors because a major process at the plant is to remove unwanted water from fresh tomatoes. Since tomatoes are about 95 percent water, a 1 percent increase in solids content has a very significant effect on the quantity of processed products produced. In addition, a reduction in the amount of water in the fruits reduces cooking time and cost. However, breeding for high solids content is particularly difficult because there is a negative correlation between yield and solids content (Stevens and Rick, 1986).

To a considerable degree, processors control the definition of quality of processed tomato products because of their particular location in the system. Most grower contracts specify the cultivar to be grown and require that the tomatoes be graded. If the processor decides to switch to a tomato cultivar with higher solids, the annual contract with growers is modified to reflect this change. In addition, processors test their products in various ways. Gould (1983:196) distinguishes two dimensions of quality. Subjective measures of quality are "based on the investigator's opinion ... Examples include flavor, odor, color, tactile [sic], or texture." In contrast, "... objective quality evaluation is based on observations that exclude the investigator's attitude. As recognized standard scientific tests, they are applicable to any sample of the product or products without reference to its previous history or ultimate use. They are representative of modern quality control because the human element is excluded" (1983:197).

While one might dispute the sharp distinction between subjective and objective measures that Gould makes, his typology is important in showing how scientized the tomato industry – as all of contemporary Western agriculture – has become. Minor differences in the results of a tomato colorimeter test can and do determine considerable price differences in the product sold to the consumer. Yet, despite great efforts to develop laboratory tests, flavor evaluation still remains the province of human testing.

With the mechanization of tomato production has come a more concentrated processing industry in the US. Brandt et al. (1978) report that, nationally, the number of processing firms declined from 424 to 118 between 1956 and 1975, a drop of 72 percent. Concomitantly, the annual tonnage processed during the same period increased from 9,600 to 69,400 tons per firm. In California, the tonnage per firm increased from 49,500 to 259,600 during that period. In a 1975 study, Jesse et al. (1975) argued that construction of small cleaning and juicing plants in the center of tomato growing areas, with shipment of the juice to existing plants for further processing, would be more economical than the existing arrangements; however, to date no such reorganization has occurred. Moreover, from the growers' point of view, these figures on industry concentration are somewhat misleading. Growers usually have few choices. In a given geographic area only one or two processors are likely to be signing contracts. In addition, processors rarely limit themselves to tomatoes as do many California growers.

Let us now turn again to the three nations that were included in our comparison above: Spain, Portugal, and Greece. Once again our analysis draws heavily on Uyeshiro (1977). As illustrated below, the exigencies of the international market are gradually creating a convergence in processing technologies used in these nations and the United States.

Spain About 16 processing plants serve the Extremadura region of Spain. (As of the time of the latest available information, no records were kept on the number of processing plants nationally.) Most of these are Spanish-owned, one by the government. About half of these have the capacity to handle more than 16 tons per hour while two can process 65 tons per hour. As of the late 1970s, most were much less automated than plants in the US, although rising labor costs were pushing owners in the direction of automation. Whole tomatoes constituted 70 percent of the production, and paste, 28 percent. But since paste is highly concentrated, 60 percent of the raw product went to make it. Only 30–35 percent of the paste and 50 percent of the whole tomatoes are consumed domestically. About half of the exports go to EEC countries. Tin cans, once made by hand during the off-season, largely by the female (two-thirds) labor force, are now purchased. The manufacturing plants have few quality-control features and little environmental control. However, since much of the waste is sold as animal feed, environmental problems are few. To take advantage of the many benefits, many plants are now diversifying into other vegetables and fruits. Finally, while a subsidy was granted to processors in 1972 and 1973, it was dropped in 1974 and 1975. However, in the late 1970s, credit continued to be provided to processors and assistance programs were available to encourage exports.

Portugal   The first paste manufacturing plant was built in Portugal in 1939 and it was supplied entirely by the fresh market surplus. With the help of Heinz, well-equipped plants that specialized in paste were built in the late 1950s. From 1957 to 1968, paste production soared from 3,000 to 160,000 tons, and quality controls were adopted from the US to ensure that most would be exported. Portugal had 29 processing plants by 1977 and because of overproduction, there is little prospect of any new plants being built. Thirteen of the plants were nationalized in 1975, but later denationalized (Bleche, 1987). Most of the existing plants have a capacity of 40–60 tons per hour. Unlike Spanish processing plants, Portuguese plants are among the most modern with respect to quality control, even surpassing American technology in some respects. Fully 97 percent of the raw product is made into paste. Not surprisingly, processors are desirous of broadening their product mix. In fact, two types of paste products are produced: double concentrate (28–30 percent solids) and triple concentrate (36–8 percent solids). Nearly all of the paste is exported, with about half going to the EEC. Most of the exports are shipped in 5-kilogram cans and some are shipped in drums. This suggests that it is used as an input in other tomato-based products in the importing nations. Until the 1970s, cans were made by hand during the off-season, as in Spain; now they are purchased. Today, there are still less than 30 factories and the industry suffers from chronic overproduction. And, as in Spain, pollution controls are lacking and regulation is not as stringent as in the US. Hand-harvesting of the produce works in favor of the processors because it generates little trash and soil. Hence, the waste is easily sold for animal feed. Finally, there is little direct assistance to processors although they are able to import equipment duty-free.

Greece   In Greece, 55 percent of the processing plants were built after 1960, but most are small by American standards. Although the largest plants handle only 50 tons per hour, most plants handle 10 tons or less. Private Greek firms handle 80 percent of the processing capacity and the rest is owned by cooperatives. Over 80 percent of the tomatoes are used for paste, with much of the rest canned whole. Of the total production, 85 percent was marked for export, 75 percent of which went to the EEC in the late 1970s. At that time there was some indication that the market might be restricted by an EEC ruling. However, capacity actually increased in expectation of entry into the Common Market. Today there are about 50 plants, half of which specialize in tomato processing. Quality remains variable, with hot break being exported in plastic bags to the United Kingdom for use by Heinz and lower quality paste being sold in the Middle East (Bleche, 1987). Break refers to the initial portion of the extraction process whereby the tomatoes are crushed and chopped. This may be done with either hot or cold tomatoes; hence hot or cold break. The Greek government intervenes in processing as it does in production. It provides rebates on the interest paid by canners for funds borrowed during the harvest season. in addition, it sets a minimum export price. One effect of the considerable state intervention in Greece has been the growth of processing plant capacity beyond the size of the crop.

The four cases described above illustrate the diversity of organizational structures and government intervention in tomato processing in California and in the three European nations, in addition to the world market's growing competition for canned tomato products. As a result of the increased competition, the Mediterranean nations formed an association called AMITOM (The International Association of Mediterranean Processing Industry) in 1980 (Bleche, 1987). In each country, the state has fostered the growth of the industry with the express aim of bringing new technology (and in some cases new organizational forms) to bear on production and processing costs. Taken together, these cases also illustrate the technical convergence that is occurring in both tomato production and processing, a convergence that will bring with it even greater competition in the future. This competition, in turn, will increase the potential profitability of the new biotechnologies for tomato production. As we shall see, these new technologies offer considerable opportunities for a relatively new set of actors that have been only marginal to tomato production and processing in the past. However, before discussing those technologies, let us examine the production and marketing of fresh market tomatoes.

Fresh Market Tomatoes

Production

Unlike processing tomatoes, most fresh market tomatoes are still picked and packaged by hand (although in California, "mature greens" for the fresh market are increasingly harvested by machine). This is a laborious, time-consuming process and there is considerable interest in developing a machine more suitable for the harvest of fresh market tomatoes (Stevens, 1986). Unlike the processing tomato industry, where most growers are not engaged in postharvest activities, the grower-shipper is the norm in fresh market production. This is undoubtedly the result of the limited economies of scale and even considerable diseconomies that would occur if packaging were removed from the point of harvest, a point that is constantly moved so as to provide year-round production.

Most fresh market tomato plants are grown from transplants that are either produced by the farmer or bought from local nurseries. In addition, since the fresh market tomatoes will be purchased by the consumer (except for those served in restaurants), they must exhibit certain characteristics which are unnecessary and/or undesirable for processing tomatoes. Tigchelaar (1986:147) explains:

> The marketplace demands large, round fruit (to fit a hamburger bun conveniently?) with adequate firmness and shelf life for shipping to distant markets; uniform fruit size, shape, and color; and freedom from external blemishes and abnormalities. These features must be combined with the required horticultural characteristics – earliness, growth habit, disease resistance, and adaptation to environment – to make a successful cultivar.

Tigchelaar further notes that there has probably been more emphasis on appearance than on flavor. This is hardly surprising since there is little brand identification in fresh market tomatoes and the consumer cannot test flavor before purchase. The consumer, unfortunately, falsely assumes that appearance is correlated with flavor. In particular, American and European consumers associate a bright red color evenly distributed throughout the tomato as a sign of ripeness and, hence, flavor.[4] Nor should Tigchelaar's offhand comment about hamburger buns be taken lightly: in 1985, McDonald's became the largest single buyer of fresh tomatoes in the United States.

Postharvest

Transportation is the largest single cost in the marketing of fresh market tomatoes (Mongelli, 1984), which is due, in part, to the great distance the produce is transported from the producing areas to the consumption areas. Today, most fresh market tomatoes in the United States are harvested green so that they are not overripe when they reach the market (figure 5.1). After harvest, they are taken to the grower-shipper packing sheds where they are washed, sorted by size and color, and usually packed in fiberboard cartons that hold 20–30 pounds of tomatoes each. Growers usually operate the packinghouses and thereby gain substantial control over the production process. Packing green tomatoes has its costs, as Mongelli (1984:18) explains: "Many mature

---

[4] In contrast, Japanese and Korean consumers prefer pink tomatoes (Villareal, 1980). In many other nations, tomatoes are consumed green.

green tomatoes are shipped to repackers in terminal markets. These repackers specialize in ripening and resorting tomatoes for color uniformity." Not surprisingly, this repacking adds significantly to the cost of production. With rising labor costs and increased packing costs, the repacking industry has begun to decline in recent years.

Greenhouse Tomatoes

In the US, greenhouse tomatoes comprise only about 5 percent of total tomato production, but in Europe, they are considerably more important. Greenhouse tomatoes are usually grown from transplants using varieties with indeterminate vines (i.e., those that continue to grow and set fruit throughout the season). Under optimal conditions, tomatoes complete their reproductive cycle in 95–115 days and, therefore, produce three crops a year when this approach is used. Greenhouse tomatoes are used exclusively for the fresh market where they usually receive higher prices than those grown in the field, especially during the winter. There are a number of reasons for this, but perhaps most important is the higher quality. Unlike field-grown fresh market tomatoes, those produced in greenhouses are picked only when fully ripe. This vine ripening of the fruit has conclusively been shown to lead to a tastier and more attractive product.

However, greenhouse production has certain characteristics which have limited its growth; these include higher capital and labor costs and greater energy consumption. A variety of problems are also created by indoor cultivation. First, tomatoes outdoors are self-pollinating. When grown in the field, the wind is sufficient to shake the pollen off the stamens and onto the pistils of each flower. In contrast, in the greenhouse, flowers must be pollinated by manually vibrating the plant. Second, the greenhouse's higher humidity is a fertile environment for several tomato diseases. Breeding to develop disease-resistant varieties as well as a stringent spray schedule is often necessary to keep diseases under control. Finally, the greenhouse environment reduces the amount of light the tomatoes receive. Although tomatoes are not photoperiod sensitive, they thrive with substantial direct sunlight.

Home Gardens

Tomatoes are planted by 90 percent of US home gardeners (Tigchelaar, 1986), as well as by home gardeners around the world. While home garden production statistics are unavailable, as a group they do certainly constitute a considerable proportion of total production. What is of greater importance to us here, however, is that home gardeners have different objectives than commercial growers and, therefore, make different demands on the plant breeders. For example, home gardeners are interested in continuous production throughout the growing season. They are not interested in determinate plants and are more than willing to invest considerable labor to trellis, prune, harvest, and otherwise care for a few tomato plants. Since home grown produce does not have to be shipped, highly perishable but flavorful varieties are popular among home gardeners. Finally, home gardeners are much more willing to experiment with novel cultivars than is the public at large. For example, seed for orange-colored varieties can find a market niche among home gardeners (Tigchelaar, 1986).

Breeding: The Red and Green of It

Tomato improvements may be divided into four periods for historical analysis. The first was characterized by the use of sports, accidentally occurring mutants noticed by farmers and others in the field. The second was characterized by breeding using Mendelian principles, a point at which the distinct processing and fresh market varieties arose. The advent of large-scale hybrid seed production marked the third period. Finally, the use of molecular and cellular techniques in the last few years constitutes the beginning of the fourth period. Let us briefly examine each of these.

Seedsmen and farmers, and occasionally scientists, made the first tomato improvements through the use of sports. One gene, known as *sp* (for self-pruning), appeared in 1914. A farmer by the name of C. D. Cooper found a plant that contained *sp* and marketed it as "Cooper Special". This gene is essential to machine-harvested tomatoes because it creates a more compact growth and concentrated flowering. This, in turn, creates a high plant density allowing a single harvest. Similarly, the first *Fusarium* wilt resistant variety, developed from a single plant found in a heavily infested field, was released in 1912 as "Louisiana Wilt Resistant" (Stevens and Rick, 1986). The use of naturally occurring mutants is still important in breeding today; over 1,200 single-gene mutants have been identified.

The second period of tomato improvement began in the late 1920s and involved the application of Mendelian principles to breeding tomatoes. Much of the initial work was conducted in the SAES and USDA laboratories (and similar government-supported institutions in other nations) as the investment was substantial and the returns did not remain with the breeder.[5] In the United States, disease resistance was a major breeding goal. Marglobe, a cross between the Marvel and Globe varieties, was developed in a USDA greenhouse by Pritchard and Porte in 1918 and released in 1925. In the *Yearbook of Agriculture*, Boswell (1937:185) noted that Marglobe "is without doubt the most important variety of tomato in the United States and the world today." Although initially developed for fresh market use for shipment from Florida north along the eastern seaboard, Marglobe's disease resistance – both to wilt and rust – soon made it a major canning variety as well. Nevertheless, certain processors, especially Campbell (see Boswell, 1937; Doyle, 1985), continued to develop their own breeding programs to ensure the high quality of the crops for which they had contracted. Such work, especially in the case of greenhouse tomatoes, has usually been conducted in secrecy (Stevens and Rick, 1986).

A key point in conventional breeding was the development of the machine-harvestable tomato by G. C. Hanna of the University of California at Davis.[6] Although most of his colleagues, as well as the growers, thought it was impossible, Hanna began his breeding work in the mid-1940s. In 1947 he toured the country for six weeks visiting all the major tomato growing areas, returning to California with seeds of varieties he felt had certain characteristics that would make them amenable to machine harvesting. Five major charateristics were necessary in new tomato varieties: the tomatoes had to develop along the entire vine at once, the fruit had to be firm, the fruit had to be easily removable from the vine, most of the fruit had to ripen at the same time, and the fruit had to break off without the stem attached, as it might otherwise puncture the machine-harvested fruit. In addition, the new cultivars had to incorporate all the qualities processors desired and had to yield about the same as the existing cultivars.

By 1958 Hanna, in collaboration with Coby Lorenzen of the Agricultural Engineering Department at Davis, had developed a prototype machine and a prototype tomato. "The new crop had a small vine, concentrated fruit set, and adequate firmness to withstand machine handling in addition to having the required yield, disease resistance, and quality characters" (Tigchelaar, 1986:144). The pear-shaped prototype tomato was found to withstand the rough handling of the harvester better than the existing round varieties (Rasmussen, 1968). Hanna then sought and received the support of Ernest Blackwelder, a manufacturer of farm equipment, and Lester Herringer, an active member of the

---

[5] As the reader will note, most of the citations to breeding work are to Americans. This is not merely due to the vastness of the literature. A recent survey estimated that 50 percent of all tomato improvement was carried out in the United States (Portas, 1987).

[6] This section draws heavily on Friedland and Barton (1975).

California Tomato Growers Association. The processors and growers were generally receptive to the idea of the harvester. Herringer offered test plots on his own fields and convinced several friends to do the same. Although the trials were not conclusive, they did demonstrate that machine harvesting was possible. However, the 1959 trials revealed that the lugs were not suitable for machine harvesting. O'Brien, also of the Davis Agricultural Engineering Department, began work on the development of a large bin system and finally came up with the idea of two tractors each pulling a trailer with four bins. These engineering changes, plus the redesigned tomato, ensured the workability of the new system.

Another development in tomato breeding has been the widespread focus on hybrids. The first large-scale use of hybird tomato seeds was in Bulgaria in 1932, while the United States and Western Europe did not employ them until after the end of the Second World War (Yordanov, 1983). Hybrids have been widely used for greenhouse tomatoes since the 1960s, although it is only more recently that they have been widely used for field-grown fresh market and processing tomatoes. Hybrids are quite expensive, with seed selling from four to 15 times that of varietal seed (Villareal, 1980). However, since seed is only 2–4 percent of the cost of production, many farmers are willing to use such seed, especially for fresh market production.

Production of hybrid tomato seed is extremely labor-intensive. First, the male portion of each tomato flower must be emasculated by hand and pollen must be collected from the male parent. Because it is difficult to store the pollen for more than two or three days, the female plants must be pollinated as rapidly as possible, creating a great demand for labor. Fifty pollinators for 30 days each, or about 12,000 hours of labor, are needed to fertilize, individually, each flower on the plants in a single hectare.

One way to reduce labor costs is to develop male-sterile lines. Although this approach has been successfully accomplished for several other crops, it has not yet been perfected for tomatoes, in spite of early breeding attempts by Rick in 1945 (Yordanov, 1983). The problem is that unless sterility approaches 100 percent, some of the resulting seeds will produce only stunted plants. Even a few such plants would be unacceptable to growers, especially given the much higher prices commanded by hybrid seed. Hence, work continues to develop true male-sterile lines. One novel approach has been to develop a marker that would identify seeds having (or not having) the desired characters (e.g., Atanassova and Georgiev, 1986; Jorgensen, 1987).

Even with the use of male-sterile lines, labor would only be reduced by 40–50 percent. Because of the high volume of labor needed, hybrid production is only feasible where labor is cheap, yet skilled, and the

climate is suitable. To date, Taiwan remains the major producer of
hybrid tomato seeds, mostly for foreign sales. In 1977 Taiwan produced
15,000 kilograms of seed, enough for 100,000 hectares of tomatoes.

Of particular note is that the use of hybrid seed does not increase
yields of tomatoes. Villareal (1980:83–4) explains:

> Common varieties now exist that have traits comparable to hy-
> brids. Nevertheless, in countries ... where the seed trade is
> independent of government control, common varieties have dis-
> appeared from seed catalogs for several reasons. The huge invest-
> ment involved in the developing of common varieties is lost within
> a few years. If a company develops a common variety, a farmer or
> a competing company can buy the seeds one year and produce
> seeds of the variety the following year. This does not happen with
> hybrid seeds because the parental lines are kept secret. Another
> reason for the rapid adoption of hybrids in developed countries is
> the extensive promotional campaigns of the seed companies.

On the other hand, Tigchelaar (1986) argues that hybrids are superior
in terms of earliness[7] and consistency of performance in suboptimal
climatic conditions.

Today, nearly all fresh market tomatoes in both the United States and
Japan are produced from hybrid seed. In the US hybrid seed for fresh
market tomatoes sells for $1.25–$2.50 per gram. About 25 percent of
US processing tomatoes are now hybrids and are produced from seed
that sells at $3.75–$4.85 per gram (Stevens and Rick, 1986). Irrespective of breeding goals, the method favored by most breeders today
involves an initial cross followed by pedigree selection (Tigchelaar,
1986). In contrast, breeding goals vary widely among tomato breeders
just as the uses vary with the crop. Yield is rarely, if ever, a breeding
goal because it is composed of many factors.[8] On the other hand, any of
these individual yield factors – perhaps best looked at as constraints
– may be the subject of a breeding program. As noted above, disease
resistance is of concern to breeders but, at least in the US, it is a
problem that is considered largely under control. Insect resistance is

---

[7] Yordanov (1983) notes the important distinction between biological earliness, defined as the time between seeding and the first ripe fruit, and economic earliness, defined
as the time between seeding and harvest. In addition, earliness for the fresh market is
desired so as to permit marketing when prices are high. Earliness for growers of processing tomatoes is defined in terms of ease of management of the harvest and delivery to the
processing plant. The price in the latter case has been already fixed (see George and Berry,
1983).

[8] The limits of breeding for yield are apparent. During the 1960s, processing tomatoes
purchased by California processors increased in net weight by 10 percent. However, all of
the increase was water content – water that later had to be removed (National Commission on Productivity, 1973).

rarely addressed as the insecticides that do exist are effective and cheap (Tigchelaar, 1986). Quality, however, is very important in breeding circles.

Nearly all breeders talk of the importance of breeding tomatoes of high quality; however, they disagree as to just what constitutes quality. George and Berry (1983) suggest the following are important to breeding quality: soluble solids, low pH, titratible acidity, and color. Villareal (1980) argues that quality requirements for processing tomatoes are clear and include high solids, low pH, firmness, ease of peeling, crack resistance, and color. This ambivalence is perhaps best expressed by Stevens (1986:569) who writes: "In most tomato-breeding programmes, fruit quality receives more lip service than actual effort. Nevertheless, there is great interest in improving the quality of tomato fruits." By at least one indicator, vitamin content, quality has improved markedly; cultivars released in the 1970s had 25 percent more vitamin C than those released in the 1950s (Stevens and Rick, 1986). However, increased vitamin C is associated with low yield and odd-shaped fruit. Moreover, "vitamin C and sugar content decrease when tomatoes are harvested at the mature green stage and ripened in storage or while in transit. In vine-ripened fruits the opposite occurs" (Villareal, 1980:79). Thus, the development of a large-scale, highly concentrated and regionally specialized fresh market tomato industry mitigates against maintenance of high levels of vitamin C. Similar problems exist with breeding for beta carotene, a precursor of vitamin A. While it can be increased through breeding, it almost always turns the fruit orange, a color largely unacceptable to most Western consumers.

Fresh market tomato quality has been the subject of considerable discussion in the United States. Since most fresh tomatoes are harvested green and shipped long distances to consumers, there is widespread public complaint about their lack of quality, in this case, lack of flavor. Tigchelaar (1986) argues that fruit quality has remained a key focus among breeders despite public perceptions to the contrary. Although we have no way of providing clear empirical data on this point, it does not appear unreasonable. The irony is that the ability to produce off-season tomatoes in California, Florida, and Mexico and ship them to New York has created expectations among consumers that winter tomatoes will taste just like summer tomatoes grown locally or even in their backyard gardens. Moreover, the same methods used to produce off-season tomatoes are now used year-round; hence, the problem of taste is no longer confined to the winter. In short, with respect to flavor, off-season production has created a new set of problems that did not exist before.

Although tomato germplasm is particularly rich, tomato breeding is constrained by the availability of satisfactory genetic material. Thanks largely to the work of Dr Charles Rick (now Professor Emeritus at the University of California at Davis), who founded and for many years nurtured the Tomato Genetic Stock Center, an abundance of wild species of tomatoes have been collected, stored, and evaluated. However, there is less variation in cultivated tomatoes than in maize (Rick, 1986)[9] for two main reasons: only a few varieties can withstand shipment (Plucknett et al., 1987), and certain single genes have maintained resistance for many years – for example, the Florida tomato industry has relied on a single gene to control wilt since 1949 (Plucknett et al., 1987). There are many wild relatives containing genetic material that confers disease resistance and other traits upon tomatoes. These wild relatives have been used in tomato improvement programs since the 1930s. In fact, "the introgression of wild traits has been so extensive during the last 45–50 years that without it, production would be much reduced, current cvs. [cultivars] would be far less useful, and tomatoes doubtlessly could not be grown in all areas they presently occupy" (Rick, 1986:45).

The New Technologies

Tomatoes have already begun to play an important and leading role in the development of new strategies for plant improvement. Both Campbell's and Heinz have begun research to use rDNA to modify the texture, taste, color, and shape of tomatoes (Goodman, Sorj, and Wilkinson, 1987). Nor are potential profits the sole reason for the interest in tomatoes. Stevens (1986:574) explains:

> In addition to its importance agriculturally, the tomato has almost every characteristic required by biologists for a model plant system: (1) there has been extensive genetic and cytological mapping of the species; (2) there are many genetic markers available; (3) both the cultivated and wild species have vigorous growth under a wide range of cultural conditions; (4) the tomato is easily manipulated in a hybridization programme; (5) generation time is relatively short, and in a well-managed programme, three generations per year, can be obtained.

Moreover, as Tigchelaar (1986) notes, tomatoes have high variability and one cross may yield several hundred seeds. These and other features of tomatoes have made them a popular model crop among scientists involved in plant biotechnology. This widespread interest in them as

---

[9] As a result of the high degree of uniformity in the field, bacterial speck has become a problem for tomato growers (Kannenberg, 1984).

a model crop was recently evidenced in the publication of an entire volume devoted to (and entitled) *Tomato Biotechnology* (Nevins and Jones, 1987).

As of 1979, more than 970 specific genes of the tomato had been identified, in part through the work of the tomato genetics cooperative established by Charles Rick in 1951. This detailed information, available for few other crops, makes possible research programs that will, if successful, have a profound effect on the tomato production and processing industries. For example, Calgene recently announced that it has developed a genetically engineered tomato that resists spoilage, retains its firmness longer, and has increased density (Russell, 1989). This improvement is only possible because the basic principles of genetic engineering of tomatoes have been sufficiently developed and the genetics of firmness and density are well understood (Kunimoto, 1986).

Genetic engineering can also be used to encourage the creation of mutants or sports, so that one does not have to wait for them to appear naturally in the field. For example, Evans et al. (1984) have noted that cell and tissue culture of gametes and leaves can yield useful genetic variation. Using these techniques, they have identified both a tangerine-colored variety and one with a jointless pedicel (important for mechanical harvesting). They estimate that one in every 25 regenerated plants is a mutant. According to the authors: "These technologies allow the breeder to leapfrog in the accomplishment of breeding objectives using existing varieties for the rapid development of new breeding lines" (Evans et al., 1984:760–1). Such procedures appear to reduce the average backcross breeding time for new varieties from seven to eight years down to four. The biotechnology company DNAP has already received a PVPA certificate for developing a high-solids tomato in this manner (Nevins, 1987).

The consensus in the field seems to be that successes in biotechnology will come first with glyphosate tolerance and then with disease resistance (Stevens, 1986). Recent developments appear to support this view. A recent article by Fillatti et al. (1987) notes that in greenhouse experiments, expression of the mutant *aroA* gene in a commercial cultivar of tomato conferred tolerance to glyphosate when sprayed with 0.84 kg of active ingredient per hectare. While these plants still do not perform as well as untreated plants, they represent substantial progress in this domain. Still another experiment (Cocking, 1986) reported successful use of protoplast fusion between tomato and *Lycopersicon peruvianum*, resulting in fertile progeny (although *L. peruvianum* is a wild relative that often produces sterile progeny when crossed sexually with tomatoes). This technique makes it possible to include genes from *L. peruvianum* in conventional breeding programs. In contrast, Koornneef et al. (1986), in similar research using conventional breeding techniques, are attempting to produce a tomato genotype that is particularly amenable to cell culture and genetic manipulation.

Currently, most new genetic material introduced into commerical cultivars for experimental purposes has used the *Agrobacterium* vector. However, the regeneration of transformed cells still remains the major barrier to widespread use of these new techniques for tomatoes as well as for other crops (Nevins, 1987).

To date the new technologies have not resulted in any major changes in the field, although breakthroughs appear imminent. Several years ago the following forecast was made: "A better processing tomato is on the way. By the late 1980s, plant tissue culture and recombinant DNA technology will result in higher solids and acid content, square shape to facilitate shipping, and uniform color" (*American Vegetable Grower*, 1984:26). While progress has been made in achieving these goals, the forecast appears to have been overly optimistic.

Regulation

The rapid growth of biotechnology strategies for tomato improvement will put the tomato industry in the forefront of many of the thorny issues surrounding food safety. As Nevins (1987:5) points out: "Those who promote tomatoes containing novel genes will be faced with a special responsibility to ... demonstrate that prudence has been exercised in the evaluation of the global impact of modified organisms." Tomatoes, as food products, are already under the jurisdiction of the FDA, which restricts labeling, adulteration, food additives, and so on. However, the new biotechnologies create new issues and questions. The current FDA rules note that "significant alterations" due to breeding may call for review and testing. For example, does the insertion of a foreign gene to increase herbicide resistance constitute a "significant alteration"? "The labeling restrictions of the Act present one of the most challenging FDA regulatory issues: When should a tomato no longer be called a tomato?" (Gibbs and Kahan, 1986:20). Clearly, these issues will only be answered as they are brought to the FDA on a case by case basis. Nevertheless, they are bound to have a profound effect on the plant improvement process, closing some doors and opening others.

Conclusions

Many of the changes in the processing tomato industry over the past several decades are similar to those affecting the wheat industry during the last century. The tomato plant has been successfully redesigned by

Table 5.1   Price spreads for fresh and canned US tomatoes, 1919–88.

Year	Fresh tomatoes[a]		Canned tomatoes	
	Retail price (¢/lb)	Farm value[b] (¢/lb)	Retail price[c] (¢/can)	Farm value[c] (¢/can)
1919			13.3	1.9
1920			12.1	1.7
1925			11.2	1.4
1930			10.2	1.4
1935			8.6	1.1
1940			7.1	1.1
1945			10.2	2.5
1950	24.4	9.3	12.6	1.8
1955	27.4	9.0	15.1	1.8
1960	31.6	11.4	15.9	1.9
1965	34.1	12.0	16.3	2.4
1970	41.9	15.0	21.3	2.6
1975	57.8	21.4	35.0	4.8
1980	67.4	27.7	42.2	4.5
1985	77.8	27.7	51.5	4.9
1986	82.4	28.9	51.4	4.8
1988			53.6	4.3

[a] Price spreads for fresh tomatoes prior to 1950 and after 1986 were unavailable.
[b] Farm value data for fresh tomatoes from 1975 to 1986 were calculated based on a loss factor of 15 percent. This is the same as the loss factor used by USDA in computing regional price spreads (see USDA, 1987b).
[c] Calculated on the basis of a no. 303 can.
Sources: USDA, 1957, 1972, 1982a, 1987, 1989

breeders to be mechanically harvested. In some nations, the state has intervened to encourage the growth and development of the industry, to provide quality standards, to regulate processing, and, of course, to support research. Also like wheat breeding, tomato breeding is a negotiated process in which various interests attempt to encourage the development of a plant that best suits their desires.

The situation in the processing tomato industry also differs considerably from that of the wheat industry. Since the number of tomato growers and processors has declined, the growers are able to negotiate contracts with the processors as a group. This negotiating ability is perhaps reflected in the fact that the growers' share of the retail price of canned tomatoes has changed far less than that of wheat growers since 1919 (table 5.1). Industrial appropriation and substitution (Goodman et al., 1987) has been less successful to date here than in the wheat industry.

The fresh tomato industry, on the other hand, is both more complex and more diversified than the processing tomato industry. While much of the US market is held by large growers in California and Florida (with substantial imports from Mexico), there is still room for smaller producers who can gain advantages through closeness to the market and/or greenhouse production. Moreover, since most fresh market growers are also packer-shippers, they have maintained a high proportion of the value added in the packing shed located on the farm, which is reflected in the fact that farm value of fresh tomatoes is approximately one-third of the total retail price (table 5.1).

Over the next several decades it is likely that the trend toward specialization in variety development will continue. Already there are clear demarcations among varieties for fresh market, greenhouse, processing, and home gardening. It is certainly possible that breeders will employ their new tools to develop varieties that contain certain very specific characteristics of interest to the processors and shippers. For processors, the possibility of developing varieties that have certain flavor or texture characteristics exists. For the producers of fresh market tomatoes, developing tomatoes that are nearly square (which would pack more into a given space) and maintain flavor even though picked green (Jones and Scott, 1983), would be particularly sought-after characters.

At the same time, as was the case for wheat a century ago, an international market for processed tomato products is developing. Taiwan, Bulgaria, Hungary, Romania, Greece, Portugal, Spain, and Italy all have significant export-oriented production and processing industries (Uyeshiro, 1977). Most have been developed only recently. Brazil (Maluf, 1986) and India (Tikoo, 1986) are beginning to enter processing tomato production in a substantial way, and other nations, undoubtedly will follow suit. Therefore, we can expect increased worldwide competition for canned tomato products and increased pressure to adopt the new products of biotechnology, perhaps before they are adequately tested. Moreover, while a system for regulating biotechnology exists in the US (although some would argue that it is far from adequate), in many nations there is no such system. Indeed, even requirements for reporting the use of genetically altered materials to a government agency are lacking in most nations. At the same time, there are few rules in any nation concerning what must be revealed about the growing or processing of imported canned products. Thus, all nations are potentially at risk if an international system of regulation is not established.

The market for fresh tomatoes is also growing in size. What used to be, by its very nature, a local product is now rapidly being internationalized. One organization that is helping this to happen is the OECD. Since 1962, the OECD has been engaged in a "Scheme for the

Application of International Standards for Fruits and Vegetables" (see OECD, 1983). This scheme, involving more than 20 nations, has as its goal the production of highly standardized fruits and vegetables. Standardization, while imposing significant constraints on producers and distributors, has at the same time contributed to integrating the markets and restructuring the distribution system, particularly by making it possible to sell high-quality produce in standardized bulk consignments which are fully guaranteed in terms of space and time. "This trend toward the marketing of goods in bulk has brought about a change in the function and structure of markets – both at the production and wholesale levels – with the development of selling of samples without seeing the actual merchandise" (OECD, 1988:14).

In the case of tomatoes, the adoption of such standards (which are now compulsory for exports from several nations) has meant that exports of fresh tomatoes are restricted to those that meet certain conditions, including appearance, shape, development, firmness of flesh, and limitations on defects and cracks (OECD, 1988). Detailed photo brochures and nine-month training courses are offered to interested parties in how to use and properly interpret the standards for tomatoes and more than 25 other crops. In addition, the packaging is standardized to permit easy palletization for shipping. As the OECD admits, such standards are not without import for the producer.

> The producer must constantly endeavor to adapt himself to market conditions, with the immediate result that any varieties of low market value will be eliminated.... The producer must also realize that today the standardization of products begins when they are still on the tree or in the field and that the result of his efforts depends on the extent to which his production is adapted to market requirements. (OECD, 1983:49)

Such standardization, of course, will require more breeding for uniformity. Together, breeding, new cultural practices, mechanization, and restructuring of markets are likely to transform tomatoes in much the same way that wheat was transformed more than a century ago.

What just a half century ago was a garden crop consumed locally is now multinational in scope. Today Mexico exports fresh tomatoes to the US. Israel exports fresh tomatoes to Europe. The Dutch have broken all yield records on greenhouse tomato production, and most of the crop is exported. Yet, there are few restrictions in any nation on what may be done to tomatoes before they are imported. What pesticides have been used? What genes have been inserted? Who will control the world market? Who will shape food products? As tomatoes become a "world crop", it will become more and more difficult to answer these questions.

6

New Forms of Culture: The International Scope of Biotechnology

Wheat and tomatoes form just two small pieces of a much larger picture. Indeed, the last several years have witnessed the beginning of a major change in world agriculture. For the first time in history, patents or patent-like protection have been accorded to plants, microorganisms, and even animals. The large agrichemical and pharmaceutical firms have bought many of the world's largest seed companies and have begun to integrate them into their larger plans. Computer technologies and robots have begun to enter into the agricultural sector, significantly changing the ways in which agriculture is managed. Finally, the new biotechnologies have opened new vistas in agriculture, making possible for the first time the engineering of improved plants and animals and promising to make the food processing industries more efficient than ever before. Together, these as yet unrealized technical changes are likely to dwarf the so-called Green Revolution of the 1970s.

In the US, farmers and the general public tend to view the new biotechnologies favorably. A recent article in the *American Vegetable Grower* is indicative: "The promises of the new developments for agriculture have been widely publicized. For example, the potential and economic impact of new varieties of crops requiring little or no input of costly fertilizer is obvious ... [as is] the development of varieties with better resistance to disease and insects" (Boyer, 1984:51). Unfortunately, our examination of the field suggests that the author's enthusiasm is at best likely to be short-lived.

In this chapter we review the impact of biotechnology on the food system with specific emphasis on Third World agriculture. Of concern here is how the new biotechnologies, as well as the social and scientific changes that they engender, are likely to be used by powerful interests to change what we eat. In the past, agriculture has differed from manufacturing in part because of its need for land and its dependence on seasonal changes. Lenin ([1907] 1938:85) noted that "agriculture possesses certain features which cannot possibly be removed (if we leave aside the extremely problematical possibility of producing albumen and foods by artificial processes)." Yet, it appears that just such a possibility is on the horizon. Is it possible that biotechnology will create a world of superabundant food? Is hunger to be finally eliminated as a part of the human condition? Or are there other forces that will shape the biotechnology revolution to other ends?

Saving the Metaphor

Research in plant biology has, until recently, rarely descended below the level of the plant organ or part. This was partly a result of low funding levels, but was also due to the greater difficulties of working with plant cells and the strong applied character of the plant sciences. Conventional plant breeding – what we may refer to as the old biotechnologies – has proceeded with some help from Mendelian genetics, but with a heavy and necessary dose of empiricism. In the last decade, however, as described in chapters 1 and 3, it has become possible to apply techniques developed for molecular biology to plant improvement. The early proponents of plant molecular genetics and genetic engineering saw few problems in transferring the techniques developed for working with microorganisms to plants. Indeed, they often perceived plants as *the same as* microorganisms. As Chaleff (1983:679) explains: "With recognition of the similarities between cultured plant cells and microorganisms came the expectation that all the extraordinary feats of genetic experimentation accomplished with microbes would soon be realized with plants. But because of the many ways in which cultured plant cells are unlike microbes, these expectations thus far have not been well-fulfilled" (see also Schaeffer and Sharpe, 1983). In short, early proponents of plant molecular genetics took their own metaphor literally: they believed that the metaphorical equivalence "plant = microbe" was actually the case.

Despite this confusion, as noted in chapter 1, plant molecular genetics has been a growth industry. Large sums of money have been invested by public authorities in nearly every Western nation and in many Third World countries. In addition, significant investments have been made by large petrochemical and pharmaceutical firms and, especially in the United States where laws are particularly favorable, by many small venture-capital firms. Moreover, since the venture-capital firms could only survive by attracting large sums of capital, they soon became the greatest defenders of the metaphor. The "hype" found in their prospectuses succeeded not only in attracting capital but also in convincing many in the public sector of the validity of the metaphor.

Although no major breakthroughs have occurred, significant progress has been made in developing plants tolerant to a widely used herbicide known as glyphosate, which is of enormous commercial potential to the companies involved. However, many of the venture-capital firms have gone bankrupt, while others have been bought by the petrochemical or pharmaceutical giants. Only a handful have been able to generate salable products, and, thereby, additional capital.

Our story does not end here. While whole plants are much more complicated and pose many more problems for molecular biologists than microorganisms, the metaphor might yet be saved if the focus of research were to change. Instead of focusing on the improvement of plants in the field, plant cells could be treated as if they were microorganisms. The techniques of growing and fermenting bacteria, already well known to certain segments of the food processing industry (especially in the production of cheese, bread, and alcohol), combined with the newer techniques of cell culture, could be used to transform the production of certain agricultural commodities into *industrial* processes. In principle, any commodity that is consumed in an undifferentiated or highly processed form could be produced in this manner. Similarly, although with greater difficulty, tissue culture techniques could be used to produce edible plant parts *in vitro*. In short, agricultural production in the field would be supplanted by cell and tissue culture factories. Lenin's ([1907] 1938) barrier to the complete elimination of agriculture and its replacement with continuous process factory production would at last be removed!

Of course, the transformation we have just described is not likely to take place within the next year or even the next decade. Indeed, it is difficult to say which research trajectories will be developed and which will be abandoned. Undoubtedly, the ways in which markets for food products and processes are organized, the regulatory requirements imposed on the industry, and the scope of patent laws will play a major role in determining just how the food system is reorganized. However, the scientific foundations for change are now being laid. Consider the following.

Markets for certain tropical commodities have already been restructured. Sugar, once an extremely important tropical commodity, has already been hard-hit by the development of corn sweeteners and sugar substitutes such as aspartame (Nutrasweet)®. The livelihoods of an estimated 8–10 million people in the Third World have been threatened (van den Doel and Junne, 1986), and it is unlikely that the sugar market will recover from its current state of overproduction and depression.

A similar change has occurred with respect to the production of cooking oils; corn and soybean oils have become major ingredients in prepared foods in developed countries. In addition, food processing techniques have been modified to make it possible to substitute oils in processing depending on relative market prices. In the United States it is now common for manufacturers to state that one or more of several oils may be used in the manufacturing process. In economic terms, the cross-elasticities of cooking oils have increased.

Although there is great variation among trees, little breeding has been done with oil palm. Oil palm, however, was one of the first crops affected by tissue culture research. In the 1960s, the multinational Unilever Corporation, owner of large oil palm plantations in Malaysia and elsewhere, realized the potential for tissue culture of oil palm (James, 1984).[1] The initial tissue culture research, conducted in Britain in greenhouses, produced high-yielding clones and increased palm oil production by 30 percent. As of 1984, 12,000 improved trees had been planted. In addition, "more uniform, shorter trees facilitate mechanization of harvesting, reducing labor costs" (van den Doel and Junne, 1986:89). Because the annual replanting requirements are about 30 million trees, there is likely to be a steady market for the clones.

Moreover, as palm oil producers now compete with producers of other oils (coconut, soy, olive, cottonseed, rapeseed, etc.), the increased production of palm oil has reduced the demand for these other oils. American soybean producers are painfully aware of this. In fact, Calgene has mounted a program to genetically engineer soybean or rapeseed so that it will produce the high-priced oils now only found in palm kernel and coconut oils (Leonard, 1987). In the Philippines, where fully 25 percent of the population is at least partially dependent on coconut production, oil production and export have dropped precipitously in recent years (Bijman et al., 1986). While estimates of labor displacement are virtually unobtainable, it is clear that displacement has been significant. "If this program [at Calgene] were to be successful, a seed bearing coconut oil (or perhaps jojoba oil) could be grown within 50 miles of Memphis and the dire effect on coconut oil prices of typhoons or the vagaries of Philippine politics would no longer exist" (Leonard, 1987:34). As more countries adopt the higher-yielding palm clones, or develop oil substitutes, it is also likely that the price of coconut oil will be depressed to near or below production cost.

Still another area of significant scientific advancement is the production of what food technologists call "fabricated foods". Fabricated foods

---

[1] In contrast, little work has been done on improving coconut oil production, in part, because most of the 60 million persons worldwide who depend on it are smallholders (Bijman et al., 1986).

(often developed from a soybean base) "differ from conventional foods in that their basic components – proteins, fats, and carbohydrates – may be derived from many sources and combined, along with the necessary micronutrients, flavors, and colors, to form an attractive product" (Stanley, 1986:65). The origin of such foods may not even be identifiable to the consumer.

Moreover, work is currently under way to produce the flavor components of expensive fragrances, spices, and flavoring agents through tissue culture. Colin and Watts (1983:731–32) explain why:

> Many of the compounds are obtained from plants that are not grown under large-scale or controlled cultivation. This instability, combined with climatic, harvesting, transport difficulties, and possible political problems in the country of origin, often leads to considerable fluctuations in the price of the compound. With each of these compounds there is interest in a more stable and easily controlled source.

In the words of one proponent of substituting industrial processes for production in the field: "The research and development effort required is well worth the effort to achieve the *in vitro* production of not only specialty biochemicals, but potentially, food, spices, and industrial commodities" (Staba, 1985:203). David Wheat (1986) reports that a number of major corporations and biotechnology companies are currently using this approach to create products such as fruit-based flavors, mint oil, quinine, and saffron. In addition, work is in progress to produce coffee, cocoa, rubber, and tea *in vitro* (Heinstein, 1985; Tsai and Kinsella, 1981; Staba, 1985; Clairmonte and Cavanaugh, 1986). Citrus pulp vesicles have also been produced *in vitro*, raising the possibility of daily production of fresh orange juice (Rogoff and Rawlins, 1987). These *in vitro*-produced flavor components would be identical biochemically to the naturally found compounds in these products; hence, they would not be artificial in the sense now currently understood, but would be true equivalents. Indeed, given the public concern over food additives in many nations, such products would be particularly marketable (see table 6.1).

Also of significance is the factory production of pharmaceuticals and industrial chemicals using tissue culture techniques (e.g., Yamada and Fujita, 1983; Dixon, 1984; Yamada, 1984; Anderson et al., 1985; Breuling et al., 1985; Misawa, 1985; Rosevear and Lambe, 1985; Zenk et al., 1988). Many of these are secondary plant metabolites and are extremely expensive, often selling for more than $2,000 per kilogram. They are used in various processes to produce dyes, astringents, pharmaceuticals [over 25 percent of prescriptions filled in the US are for drugs derived from plants (Balandrin et al., 1985)], and other rare but

Table 6.1   Markets for selected plant products

Compound	Use	Wholesale price ($/kg)	Estimated world demand ($m)
Vinblastine	Leukemia	5000	18–20[a]
Jasmine	Flavor, fragrance	5000	0.5[a]
Sapota Caranthus	Chicle	5000	18–20[a]
Lithospermum	Shikonin	4500	
Digitalis	Heart disorders	3000	20–55[a]
Rose Otto	Rose Oil	2800	12
Ajmalicine	Circulatory problems	1500	5.25
Codeine	Sedative	650	50[a]
Papaver	Codeine, etc.	650	50[a]
Pyrethrins	Insecticide	300	20[a]
Buchu	Buchu Oil	220	153
Cinnamon	Flavor	195	3.2
Quinine	Malaria, flavor	100	5–10[a]
Ginger	Flavor	100	33
Spearmint	Flavor, fragrance	30	85–90
Cinchona	Quinine	—	20–55[a]
Coffee	Beverage	12	2210
Cocoa	Beverage, flavor	4	981
Tea	Beverage	2	2917
Rubber	Tires, etc.	1	3565

[a] US market only.

Sources: Kenney et al., 1984; Curtin, 1983; FAO, 1985; Colin and Watts, 1983

essential industrial or consumer goods. Many scientists specifically link *in vitro* production with political instabilities (Colin and Watts, 1983; Fontanel and Tabata, 1987; Ilker, 1987; Zenk et al., 1988). It is clear that at least some scientists understand the ethical dilemmas posed by the use of such techniques (see Box*).

* *Applications of Plant Cell and Tissue Culture*, Ciba Foundation Symposium 137. Copyright Ciba foundation, 1988. Reprinted by permission of John Wiley & Sons, Ltd.

Riley: Chairman, a philosophical question has occurred to me in listening to the papers that have been given on the production of agricultural products from cells in culture. At the present phase of world history, we have an enormous surplus capacity to produce plant materials from agriculture. Is it not more sensible to produce plant products from plants rather than from cells in culture? On the one hand in the developed world we are seeking ways of keeping the agricultural industry active, but from the discussion here it seems that we are finding other ways of further diminishing demands on that industry.

Galun: Unfortunately, most of the exotic products cannot be produced in those countries that have agricultural surpluses. Vanilla is very difficult to grow, for instance, in Europe and the same may be true for others. On the other hand, I believe that the future of rural communities in general lies not in improving their agricultural productivity – in that way they will never achieve a really nice standard of living – but in the introduction of centres of industry and services. This would be the only solution because the efficiency of agriculture is such that prices are low. In principle, this is part of the argument of this symposium; I don't think that farmers should stay with only farming, otherwise they will never reach a high standard of living.

Riley: Yes, but of the examples that you raise, take vanilla. Despite what you say about cheating in Europe, vanilla is a principal export of, for example, Madagascar. Madagascar is desperately in need of earning foreign currency, why deprive it of that opportunity?

Galun: I don't know, people will never be willing to pay the price. The price difference between chemically produced vanilla and the natural product is about 500-fold, so companies will aways cheat.

Withers: I am very sympathetic to Sir Ralph's point about the economies of developing countries. We are not talking simply about not elevating their economies, but about the risk of actually damaging them. Another example, comparable to vanilla, is the possibility of producing cocoa *in vitro*, which has been explored in the United States.

Barz: This is certainly a very complex question to which there are several answers. One point that is quite important for certain countries is that phytopharmaceuticals sometimes have to be imported from countries which are politically unstable or unreliable. I think there is a certain interest in technologically developed countries to have a constant supply of important products.

Fowler: On this point, we recently had a discussion with the Ministry of Agriculture, Fisheries and Foods in the UK about a research programme concerned with imported food additives. One of the key features that would lead them to fund the programme was a possible reduction in the UK import bill.

On the other hand, Balandrin et al. (1985:1158) simply note:

As the natural habitats for wild plants disappear and environmental and political instabilities make it difficult to acquire plant derived chemicals, it may become necessary to develop alternative sources for important plant products. There has been considerable interest in plant cell culture as a potential alternative to traditional agriculture for the industrial production of secondary plant metabolites. This has given rise to considerable research in Japan, West Germany, and Canada.

Misawa (1985:60), making a similar case, notes the similarity between tissue culture and microbial fermentations and states that such *in vitro* techniques permit production "in any place or season". According to Misawa, four types of pharmaceutical products are currently of interest: alkaloids, steroids, terpenoids as anti-tumor compounds, and quinones as drugs against heart disease. Systems of up to 20,000 litres − large by scientific, but still small by industrial standards − have been built (Fowler, 1984), but production costs are still too high to use them with foodstuffs.

Germany and Japan have also been interested in developing tissue-culture-based manufacturing plants. To date, large scale production of seven compounds is in the development phase in these two nations (Fontanel and Tabata, 1987). Balandrin et al. (1985) estimate that the processes involved are economical when cultures produce one gram of the desired compound per liter of culture, and the selling price is between $500 and $1,000 per kilogram, although many scientists would privately put this figure either much higher or much lower. Among the technical advantages of culture techniques are achievement of standard growth conditions, utilization of plant materials that are less complexly organized than the entire plant, attainment of higher compound levels, and ease of purifying the desired product (Anderson et al., 1985). Moreover, the heterogeneity of cultured plant cells permits selection and multiplication of those that are highest-yielding. In addition, culture techniques offer the potential for creating whole new products, either through the isolation of cells producing new, naturally occurring compounds or through the manipulation of cells so as to induce the production of entirely new compounds.

Consider the claim made by Yamada and Fujita (1983:726) for a dye/astringent currently being produced in Japan: "Compared with shikon requiring 2–3 years for harvest of the plant, cultured cells permit harvesting within about three weeks,'thereby greatly shortening the production period. The content of the shikonin derivatives in the cultured cells was about 14 percent, which was extremely high compared with 1–2 percent in the normal field grown shikon" (see also Fujita, 1988). Shikonin is now selling for over $4,000 per kilogram. One ironic note is that the *in vitro* production may actually drive the price down too rapidly, given the fact that the market is small and easy to saturate (Curtin, 1983).

Although we cannot confirm or disconfirm the assertion that at least one US company has mounted a program to produce tomato pulp *in vitro*, persistent rumors exist within the genetic engineering community. This pulp would be used to produce an array of canned tomato products including sauce, paste, catsup, and purée. Similarly, it is rumored that a French company has already produced apple sauce *in vitro* and is now attempting to scale-up the process for industrial production.

As noted above, the final stage of biotechnology involves the replacement of agricultural processes with industrial processes. Here, too, much initial work has been accomplished. Cotton fibers have been grown directly from cotton cells. Unlike those produced in the field, test tube fibers can grow from both ends. Although it is unlikely that field production will be replaced in the near future, the potential is there (Board on Science and Technology for International Development [BOSTID], 1986).[2]

Similarly, the Japan Salt and Tobacco Public Corporation, and Plant Science Limited of the United Kingdom are both investigating the production of tobacco biomass for cigarettes *in vitro* (Curtin, 1983). So far the processes have proved too expensive for commercial use. However, Zenk et al. (1988) estimate that as many as a dozen plant-cell fermentation systems will be operating by the turn of the century.

Perhaps the most far-reaching proposal to replace agriculture came from Rogoff and Rawlins (1987) who suggested a scheme in which most fields would be planted with biomass perennial crops, harvested as needed, and turned into sugar solution using enzymes. The sugar solution would then be piped to factories in metropolitan areas, and used as a medium and nutrient source to produce food through massively scaled-up tissue cultures. "In this scenario, edible products are defined as foods synthesized from separately manufactured food components, or as plant organs, plant parts or their derivatives, produced as such and not derived from a whole plant grown in soil" (Rogoff and Raw-

---

[2] Wittwer (1986) has noted that tissue culture of pyrethrin has made it possible once again to grow this insecticidal plant in the field more cheaply than to synthesize the insecticidal chemical in a laboratory. This should give us cause to withhold any simplistic generalizations about the role of tissue culture in replacing crops in the field.

lins, 1987:7). Under this system, processing would be a year-round activity, with only what was actually needed being produced on a given day. Canning and freezing, as well as most spoilage, would be largely eliminated, since food would be produced and consumed in the same general vicinity. In fact, the same production facility could be shifted from the production of one commodity to another in response to changing demand.

Rogoff and Rawlins argue that such a system would have other added benefits in terms of reduced need for monocropping, less soil erosion, less use of agrichemicals, reduced energy inputs, and minimal transportation costs. They go on to note that the work necessary to realize this scenario is already underway but that it is currently uncoordinated. Finally, they suggest ways to implement a more coordinated strategy.[3]

Production costs are the major stumbling blocks to such a transformation of the world food system (Zenk, 1978; Berlin, 1986). In particular, the sugar solution would have to be extremely inexpensive to produce. Yet, Rogoff and Rawlins (1987) note that postharvest transport costs would decline so much that production costs could double without significantly altering the final price. Moreover, it is unlikely that the switch to such a system would take place all at once. It takes little imagination to realize that *in vitro* production costs are likely to become competitive for those products which are either luxury goods available only in limited quantities, or heavy goods produced far from the point of consumption. In both instances, cell and tissue culture techniques offer significant advantages. On the other hand, there are forces that will inhibit the rapid development of *in vitro* production. In particular, it appears likely that large, multinational enterprises will resist the rapid demise of lucrative markets that they now dominate. In these instances, they will intervene by defensive patenting, by purchase of potential biotechnological competitors, or through government regulation of the competition. Indeed, Byé and Mounier (1984) have suggested that this is precisely why the otherwise apparently profitable *in vitro* production of proteins (largely for animal feed) has not become viable.

Since Third World nations rarely hold significant shares of multinational companies, those of their products that cannot be produced in the field in more developed countries are likely to become the targets of biotechnological research. In addition, it should be noted that Third

---

[3] Whether such a transformation would be embraced by either consumers or producers is questionable. For a critique of Rogoff and Rawlins's proposal, see Orr (1988).

World countries are much more vulnerable to these production shifts than countries in the First World or the Second World.

From Instruments to Production Techniques

Before examining the probable impacts that such changes might have, it is important to clarify the central role that science and technology play here. Traditionally, sociologists, philosophers, and historians of science have viewed science as a quest for new ideas and have paid scant attention to the central importance that instruments play in modern science (cf., Idhe, 1979; Busch, 1984; Laudan, 1984; Latour, 1987). Yet, it is obvious to even the most casual observer of a scientific laboratory that instruments are central to science. Moreover, like the rest of us, scientists – even those with ample support – always work under fiscal constraints. As a result, they continually seek new ways to reduce the costs of their work so they can accomplish more with the same amount of money. In this respect, then, scientists are much like capitalists; although they seek "credit" (profits?) in the form of publications and citations, rather than money, they are inexorably driven to reduce the cost of *scientific* production as surely as the capitalist is driven to reduce the cost of the production of goods and services.

When this argument is applied to plant molecular biology, we can immediately see its importance: plant molecular geneticists are interested in understanding the ways in which plant genetic material is structured. Even if they have little or no interest in the commercial opportunities presented by improved seeds (or other plant materials), they need to have rapid, efficient methods for moving from individual protoplasts or cells to tissues to whole plants, in order to ensure that what is created at the molecular or cellular level is expressed in the whole plant. It must be remembered that the various techniques for plant improvement described above are nested, as illustrated in figure 6.1. Recombinant techniques may be used to transfer a specific gene to a plant cell. That cell must then be cultured into a tissue. The tissue in turn must be differentiated into a plantlet (excluding for the moment the possibility of *in vitro* production). The plant must be tested in the field to see if the trait incorporated at the molecular level is expressed in a field setting. Finally, the new cultivar must be multiplied and released. If any step in this process is impossible or difficult, it will become a bottleneck to the whole process.

This procedure creates an enormous incentive to improve culturing processes. A recent article, appropriately entitled "Assembly line plants take root" provides an illustrative example:

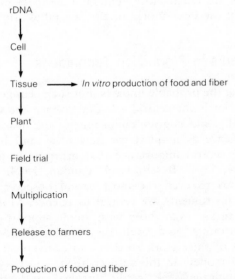

Figure 6.1   Steps involved in plant improvement and production using conventional techniques and the new biotechnologies.

The biggest thing keeping costs up – and therefore holding tissue culture propagation back – is the amount of labor needed to run a tissue-culture lab. . . . Automation is one solution to the problem of high labor costs. ARS scientists in California have found a way to automate their tissue-culture work and at the same time achieve at least two to four times the growth rate of traditional tissue-culture systems. Plant geneticist Brent Tisserat and chemist Carl E. Vandercook use a computer to control the flow of liquid nutrients to plantlets. (Comis and Wood, 1986:10)

In addition, Daly (1985) has noted that automated DNA synthesizers are already in use in both academia and industry. Similar efforts are underway elsewhere to make the plant improvement process more efficient. In short, the development of scientific techniques and equipment for use in improving conventional agriculture makes possible the elimination of agriculture as we know it. The metaphor plants = microbes can be saved by restructuring the process of plant improvement so that the culturing process produces food or fiber as the final product; then, the field, as a location for food and fiber production, ceases to exist. As one scientist put it: "We have to stop thinking of these things as plant cells, and start thinking of them as new microorganisms, with all the potential that implies" (quoted in Curtin, 1983:657).

Global Consequences

There is little question that even moderate success in realizing the metaphor would have profound effects around the world. Most immediate and most important would be the restructuring of global markets. The Third World might be able to counter this biotechnological offensive by developing its own scientific capabilities but even this would be likely to have only a limited effect. Political instability, already a problem in the developing world, would doubtless increase. Major questions about nutrition and food safety for all of us would also arise.

Market Restructuring

Even if just a few of the research programs described above were to be brought to fruition, major restructurings would occur. In the US and Western Europe, the effects on the farm would be significant but relatively minor, as the farm population is already quite small and other employment opportunities would exist for at least some of the displaced. However, it is well known that many Third World countries are dependent on one or two agricultural commodities for their continued viability. For them, the market restructurings likely to be caused by significant *in vitro* production and/or the collapse of existing markets would be enormous. A significant number of farmers and farmworkers would find themselves with no products to sell. Moreover, in both the developed and developing worlds, many of the traditional agribusiness giants might discover that their huge investments were no longer profitable. Similar effects could be expected for other tropical commodities. Let us examine the impact of a potential biotechnological replacement for one major tropical commodity: rubber.

Seventy-five percent of the world's rubber comes from Malaysia and Indonesia (Chamala, 1985). In Malaysia, 500,000 smallholders are directly engaged in rubber production while 3 million persons – approximately one-quarter of the total population – are directly or indirectly dependent on it. In Indonesia, there are an estimated 8 million people dependent on it. In both countries smallholder rubber producers are among the poorest. Worldwide, it is estimated that 22 million people depend on the rubber industry. If *in vitro* production of rubber or the use of tissue culture to improve guayule (which grows in temperate climates) (Radin et al., 1982; Fisher, 1986) were to become a reality,

90–5 percent of these people would find themselves without a cash crop (assuming the other 5–10 percent would continue to grow rubber for domestic and specialized uses). Given that the unemployment rates in nations such as Malaysia and Indonesia are already high, it would be unlikely that more than 20 percent of those displaced (about 4 million people) would find alternative employment. An estimated 16 million people, a conservative estimate, with few or no marketable skills would be without work or reduced to bare subsistence farming.

Moreover, while markets may be protected from product competition, there is little or no protection from new technology (Vergopoulos, 1985). Introduction of *in vitro* and other substitutes may result in the rapid demise of existing commodity markets, as processors restructure their production toward the new input. Without some international system of regulation and compensation, the effects on particular countries would likely be just short of catastrophic.

The degree of impact can also be estimated by looking at past instances of market collapse. For example, Mexico used to grow large quantities of barbasco, a plant from which steriods can be produced. Today, the large pharmaceutical companies have developed a method of chemically producing it, eliminating the need for the plant (Dembo et al., 1985). In the more distant past is the demise of India's indigo industry. In 1897, 574,000 hectares of indigo were grown. By 1920, with the successful development of a chemically produced indigo, field production had all but ceased. The planters attempted but failed to develop a research fund to compete with chemically produced indigo (Martin-Leake, 1975). The result, for that region of India, was widespread depression and unemployment (Kenney et al., 1984).

Moreover, displaced farmers and farmworkers migrate to urban areas in search of work. Often men leave their wives and families behind to engage in subsistance farming. If the scenario projected here proves accurate, we can expect the same pattern on a massive scale. The impact could be particularly hard on women and children since they are likely to bear the full brunt of market restructuring.

It should also be remembered from the cases of wheat and tomatoes described above that market restructuring affects not only the producers of the primary product but also those engaged in the transport, processing, and retailing of the final product. Thus, cumulative effects would likely be much greater than would be the case if changes were somehow limited to farm production. Among the possible effects are geographic shifts in the production process location, the demise and/or growth of secondary industries, the development of secondary effects as the former producing countries find themselves unable to afford imported manufactured goods, and the decline of certain consumer goods industries in the Third World as demand declines.

Finally, a significant effect of the new agricultural biotechnologies may be a shift in the geographic location of agricultural production (or its replacements) from the Third World to the First World. This, in turn, could create an increase in the already high Third World debt and a concommitant deficit in the balance of payments in Third World countries. The effects of the biotechnologies are also likely to be felt in the West. Major bank failures are possible and can only be exacerbated by market restructuring. Moreover, we can safely assume that the major banks will not be allowed to fail, but will be bailed out by the national banks of the Western powers. This, in turn, will tend to raise taxes and national debts in the West.

Third World Science

In recent years, scientists in some Third World nations – for example, India and Brazil – have succeeded in establishing significant agricultural scientific communities. Recently, these scientists have begun to develop products that can make a substantial difference in their respective agricultural economies. The International Agricultural Research Centers (IARCs) and bilateral aid agreements have provided training and support for these new scientific communities.

Nevertheless, in the past, Third World science has been derivative in character (Chatelin, 1986). Rather than entering into uncharted areas, it has been content to take care of the details in the well-charted parts of nature (Goonatilake, 1982). It has avoided the so-called "basic sciences" on the grounds – usually supported by First World donors – that all basic science was an expensive luxury. "Original scientists in underdeveloped countries work in a social vacuum, and their activity is considered perhaps ornamental, but always useless" (Goldstein, 1986:6). Chatelin and Arvanitis (1984) described the Third World as having been characterized by "scientific domination".

As a result, few scientists in Third World scientific communities are capable of pursuing the paths opened by the new biotechnologies. Only a few Third World nations have the critical mass of scientists necessary to engage in genetic engineering, although a somewhat larger number have developed tissue culture techniques for selected crops. None will be able to mount the broad research campaign needed even to stay abreast of the First World countries. One effort to do so is the International Center for Genetic Engineering and Biotechnology, supported by the United Nations Industrial Development Organization and located in Trieste, Italy, and New Delhi, India. While the Center might help in some small ways, it is unlikely that it will ever be able to take on the role played by the IARCs in supporting the research that led to the Green Revolution.

The rapid worldwide growth of technical personnel in the new biotechnologies is also worthy of note. For example, in 1931, only two persons were working in the area of plant tissue culture; now there are more than 10,000 persons (Gautheret, 1983), nearly all in developed nations. Undoubtedly, the figures for plant molecular biologists are even more skewed; salaries are high but training programs are still relatively few. Only a handful of developing countries will be able to afford to create a critical mass of scientists in this area in the near future.

Access to final products and to the processes used to produce them is another concern. The new technologies generated by the Green Revolution were created in the public sector and access to them was relatively unproblematic. In contrast, much current biotechnological research is shrouded in secrecy or its use is restricted by patents. This is particularly true in the US. Curtin (1983:649) stated: "Companies tend to be more close-mouthed about their activities in plant tissue culture than they are about culturing mammalian cells, especially in the US. Elsewhere – in Western Europe and, most notably, in Japan – their interest is more substantial and more obvious." Even Europeans are concerned about the potential consequences of being cut off from US biotechnology research (EEC, 1986). This secrecy means that technical change may appear with shocking immediacy as once stable markets collapse and market shares are suddenly rearranged.

Moreover, the Green Revolution was spatially limited by the very nature of the crops involved (Buttel and Barker, 1985). The products of the new biotechnologies will not be spatially limited even if factory production of food remains an elusive goal for some. Recombinant and tissue culture techniques can be employed, at least in principle and to a growing extent in practice, to all crops. Moreover, these same techniques can be used to extend the growing regions of crops, to increase crop substitutability, and/or to develop crops that have "functional properties" (Moshy, 1986) that make them particularly amenable to certain processing, packaging, nutritional, or esthetic uses. Bijman et al. (1986:22) argue that "those farms and regions which already have a technological advantage will be the first to benefit from the application of biotechnology. The gap between large and small, rich and poor, and between North and South will thus be widened."

In addition, even if the Third World nations are able to develop a cadre of trained scientists, they must also overcome the barriers currently restricting access to the instruments, supplies, and materials needed to do biotechnological research. Without this, at best, they can copy some of the less imaginative efforts of the West. David Baltimore (1982:33) explains:

However, ... training and the ability to self-define problems were coupled to an infrastructure of supply companies that provided specialized equipment such as centrifuges; enzymes, which are the heart of the technology; fine chemicals; laboratory gadgetry, some of which is crucial to research; specialized chromatography media; and so on. The infrastructure is possibly the most crucial part of the system.

Moreover, no research group can manufacture all or even most of the chemicals it needs. It must depend on commercial suppliers that are remote from the research site. Many of these materials – especially the enzymes – are fragile, must be shipped under special conditions, and must be delivered within several days. For example, in 1985, one large US supplier of enzymes, the Sigma-Aldrich Corporation, was able to sell $215 million in chemicals, almost entirely by mail. The company reports that 97 percent of its orders are shipped within 24 hours of receipt (Simon, 1986). Most Third World countries will have great difficulty in linking their scientists to such a supply system.

Let us assume, however, that the Third World is successful in mounting the expertise necessary to develop its own biotechnology research competence. This might be of some minimal help to offset the restructuring or collapse of various commodity markets. In some cases, patents developed in the Third World might even prevent developed countries from capturing certain markets. Tissue culture might be used to reduce production costs of field-grown crops, thereby making competition from *in vitro* production that much more difficult. However, while this might slow down the changes, it would in no way prevent the loss of millions of jobs in agricultural production. In short, the development of Third World science is no solution to the destabilizing effects of biotechnology.

Political Stability

Thus, the use of the new biotechnologies in the production of food will very likely increase political instability. In the Third World, it is likely to dash the hopes of many who yearn for a better life. While some persons would undoubtedly revert to food crop production on land now used for cash crops, popular upheavals would be likely as displaced populations demanded that their fragile states provide food and shelter. In the First World, it may also create political instability as a result of bank failures, higher taxes, and the elimination of certain processing and transport industries. In both the First and Third Worlds, it is likely that the changes forecast here will create additional social stratification, a greater gap between the rich and the poor, and a greater possibility of class conflict.

Food Security

The new biotechnologies are also likely to affect both nutrition and food safety. For those who are displaced, the specter of hunger may loom ever larger. Unfortunately, many scientists fail to realize that the production of more food does not necessarily lead to the elimination of hunger, despite ample evidence of the growing homeless population in the US. Hunger can only be eliminated when everyone has the where-withal to obtain the means of subsistence, either through access to land on which to grow food or money to purchase it (Busch and Lacy, 1984).

The problem is somewhat different for those who do have enough to eat than for those who become consumers of food modified by the new biotechnologies or produced *in vitro*. We now know that human nutrition depends on the daily ingestion of certain nutrients, although some disagreement exists as to their precise amounts. In general, we can say that we need to consume certain quantities of some micronutrients, but we have greater difficulty stating the quantities of the macronutrients. Considerable disagreement exists as to the recommended daily allowances for fats and carbohydrates, because of the difficulty of conducting tests on human beings and of the complex interdependence of the nutrients. We know even less about the interactions among foods or between foods and adulterants (Gussow, 1984).

This results in a great deal of uncertainty. Every time that we eat fresh foods we also ingest unknown quantities of the plant's natural flora including bacteria and fungi. Do these organisms play a role in our diet? Every plant product that we eat is itself the result of the peculiar circumstances that exist in a particular field. These circumstances, in turn, affect the chemical composition, flavor, and texture of the things we eat. Normally, we eat a variety of foods whose nutrient contents result from different field conditions. Moreover, we normally ingest small quantities of substances that are toxic in large quantities. We know little about the role these compounds play in nutrition. As Richard Ronk (1986:30–1) has suggested, even "increasing the amount of an essential nutrient in a particular food may be beneficial in theory, but not in practice if it predisposes the food to growth of either pathogenic or spoilage bacteria, or if it interferes with another nutrient." Food products that incorporate genetic material from exotic sources or that entirely eliminate toxic substances, especially food products produced *in vitro*, would reduce the diverse range of ingredients in our food supply and perhaps eliminate many of them. Yet, it is highly questionable whether we have the necessary knowledge to artificially construct a signficant portion of our food supply. Would such a food supply be adequate, not only in the short run, but also in the long run? What effects might it have after several generations?

Unfortunately, the present state of science fails to answer these questions. Indeed, it appears that they are very difficult if not impossible to raise. Consider the following:

1   There is a wide chasm between the area of food and nutrition and that of agricultural production (Randolph and Sachs, 1981). This gap extends from the US National Academy of Sciences, where food and nutrition are under one board and agriculture is under another, to the structure of most universities throughout the world, where colleges of agriculture are often separate from schools of nutrition. Friedland (1985) noted the lack of nutrition research on tomatoes as well as the marked tendency for graduate programs in nutrition to be located outside American colleges of agriculture.

2   Both the agricultural and nutrition sciences tend toward reductionism. That is, they tend to assume that the world can be subdivided into a series of discrete problems that can be solved serially. Without a doubt, for some problems, this type of approach can and does work quite well. However, for other problems such as those raised here, reductionist strategies fail to even grasp the problem.

3   Even now, the testing of new food products and food additives is a complex and arduous task with which few governments are able to cope adequately. The production of food and drug products, with the aid of the new biotechnologies, will undoubtedly put added stress on a system that is at best a weak one. Will these products be considered "the same" as food produced in traditional ways and so escape government testing? Or will they be considered "new" and require elaborate testing methods? (We should note here, in passing, that they are likely to be considered "new" from the standpoint of obtaining patents.) If they are to be tested, how stringent should these tests be? Indeed, can tests be devised for potential problems that are chronic rather than acute?

4   The emphasis in both nutrition and food science is not on the development of informed policies but on the clinical diagnosis of problems in the former and the development of new food products and processes in the latter. Members of both sciences often studiously avoid food policy questions on the grounds that such questions lie outside their domains. Yet, the very products that they develop *de facto* create a food policy. And, who has the expertise in these areas if not those in the scientific community?

5   An additional problem arises in the social mechanisms necessary

to maintain ever more complex food systems. Even today there are occasional documented cases of food product contamination, either by accident, by a misguided or deranged individual, or even for political reasons. The recent destruction of huge quantities of Chilean fruit as a result of a few poisoned grapes is a case in point. However, the production of food or food ingredients *in vitro* creates much greater security problems. Contamination of large quantities of food – for whatever reason – would be easier under conditions of *in vitro* production, simply because so much food would be produced in so small an area. Moreover, protection of such production facilities would be difficult and would necessitate undemocratic institutional forms. Given that some observers have encouraged *in vitro* production as a remedy for political turmoil, this problem presents both ironies and contradictions.

Conclusions

There is a certain irony in that the technological optimism described above is associated with social pessimism. However, as we have noted, even technological optimism must be examined carefully. This is – despite impressions to the contrary – an extremely difficult task because the only source of information about future directions in biotechnological research is the scientific community itself. That community is strongly biased in the direction of technological optimism. For example, some years ago scientists were infatuated with the prospect of modifying plant materials through irradiation; a little later we were told that the sea would provide the abundant food needed for future generations. Neither of these developments has come to pass. Current forecasts for *in vitro* production pay little attention to the intricate and expensive basic science that must still be conducted to modify even the most simple plant characters (including unraveling the plant nutrition process, and the relationship between plant parts and their metabolites), the complex developmental research necessary to arrive at production prototypes that have high productivity (Fowler, 1985), the difficulties of scale-up (e.g., Senior, 1986), and the economics of large-scale production (Goldstein, 1985). They also ignore regulatory barriers that others may throw in their path. For example, the FDA recently announced that only vanilla from the pod of a vanilla plant can be described as natural (Fowler, 1988). This will undoubtedly discourage corporate investment in *in vitro* production designed to draw on the public's desire for "natural" foods. All of these are likely to be formidable barriers. However, even though the scientists' prognostications may be biased, we ignore them at our peril. The possibility exists that a radical restructuring of agriculture could be triggered by powerful interests using these

Figure 6.2 The superposition of sunflowers, test tubes, and researchers on each other is used by this company to demonstrate to buyers that it is in the forefront of technical change.

(© Seedtec International. Reproduced by permission.)

technologies. We must examine the likely consequences now and ask whether we wish to pursue this path. Yet, at the same time, we must remain skeptical.

Fortunately, the new biotechnologies offer us an opportunity to assess new technology before it actually exists. This is particularly desirable in that it permits – at least in principle – both the avoidance of deleterious effects of proposed technical change through market reorganization and the creation of governmental and intergovernmental policies to direct that change. Nevertheless, much needs to be done. A more careful review and monitoring of the scientific literature is essential and more detailed forecasts of the impact of particular technical changes are needed.

Nevertheless, we can make some fairly strong statements about impacts even if we accept the "optimistic" position that some scientists take. First, we can expect a major transformation of agriculture over the next quarter century. Specifically, we can expect the not-so-gradual reduction of spatial, temporal, and climatic barriers to food and fiber production. This change alone will bring with it substantial social upheavals as the location of production changes. In addition, we can expect the elimination of major portions of the farming enterprise if field crops are grown *in vitro*. This, in turn, will displace farmers and farmworkers on a scale never before possible. Moreover, most of these people will be unable to find work in other sectors of the world economy. Thus, the demise of agriculture would deprive many people of their very means of subsistence. Another aspect of the new biotechnologies is the enormous concentration of economic and therefore political power that is likely to become possible. Already, linkages between the chemical, pharmaceutical, seed, and food industries have been formed. Finally, concentrated production would also bring with it the possibility of deliberate or accidental contamination of the food supply; in a word, it would actually reduce our food security.

The problems and issues discussed above raise complex questions of both ethics and policy. These two topics form the substance of chapters seven and eight, respectively.

7

Second Nature:
The Demand for Accountability

We have discussed a wide variety of issues, orientations, and alternatives involved in the use of the new biotechnologies in agricultural research. It should be clear from our analyses that the use of biotechnology and the concomitant changes in the law, in the roles of public and private sector institutions, and in the practice of science in general have important implications not only for agriculture but for the larger society. These implications are not merely domestic but global in nature. In this chapter, we summarize the findings of our study with emphasis on the impacts that biotechnology is and will be having on agricultural research, agriculture, and society. Clearly, from the preceding chapter it makes little sense to use the term "impact" simply to mean the effects that the new biotechnologies will cause. Instead, as we have amply demonstrated, the new biotechnologies and traditional plant breeding are part of a complex web of social relations. Although the new biotechnologies are a new element in that web, they are by no means the only new one.

As we argued in chapters 1 and 2, it is useful to take an "economic" view of the new biotechnologies, a view that sees them as one facet of a series of interrelated changes that together are restructuring the laboratory and the larger society. The *demand* for the new biotechnologies has emerged out of changes in patent law, the decreased funding of SAESs, the interests of chemical and pharmaceutical companies in linking seeds and chemicals, and perhaps the desires of (some) farmers to compete more effectively on world markets. The *supply* of the new biotechnologies has emerged out of breakthroughs in molecular biology, the availability of venture capital as a result of tax laws, and the creation of a host of new institutional forms that link previously separate research organizations. Thus, it makes little sense to talk of the new biotechnologies independently of the social, political, economic, and scientific networks of which they are a part. It is only in this setting, within this complex network of social relations in which all actors are involved in continuing negotiation, persuasion, and coercion, that we can talk of impacts.

However, since our task here is not simply to write another history of these events, it is more important that we face the crucial questions these actual and projected impacts inevitably raise: What do they mean for scientists, administrators, corporate executives, and politicians? What must they mean for us as scientists and citizens? What responsibilities do they thus entail? In this, we pass from the largely descriptive arena of social science to the normative arena of social ethics.

The Social Impacts of Plant Biotechnology

This study, like others recently published (Yoxen, 1983; Kloppenberg, 1984; Kenney, 1986), suggests that over and above technical changes the new biotechnologies are likely to generate or accelerate a number of larger social changes. Five of these changes will be discussed more fully.

A Shift of Research Emphases

The new biotechnologies appear to have accelerated a shift − which began in the 1920s − of varietal breeding from the public to the private sector. Until 1923, USDA regularly distributed free seeds, collected from around the world, to farmers who wanted to test them on US soils. As the techniques and effectiveness of plant breeding improved and seed industry pressure increased, the distribution of free seeds stopped. Instead, SAESs began to disseminate new varieties developed by public sector breeders to the seed companies which, in turn, multiplied and sold them to farmers. The development of hybrid corn shifted the division of labor again. Experiment stations gradually shifted to the development of parent material and significantly reduced the production of finished products for hybrid crops. With the growth of the seed development (as opposed to multiplication) industry came increased pressure on the public sector to cease producing finished varieties, a practice that the American Seed Trade Association (ASTA) had begun to encourage as early as 1923. Most of the pressure seems to have come from the larger companies. A recent survey of ASTA members reveals that small− and medium−sized companies place more importance on hybrid and variety development by USDA/ARS than do large companies. Similar results were found for companies that did not carry out research and development as compared with companies that did (ASTA, 1984).

From the point of view of the large seed companies, experiment stations should do basic research and produce "enhanced germplasm" that they (the seed companies) can develop into new varieties. The new biotechnologies are viewed as the tools for conducting such basic research. Both views are frequently echoed by experiment station directors. Pressure from industry to get out of varietal breeding, pressure from administrators for greater productivity, and the lure of large sums of private money for biotechnology research have led to a change in disciplines in the SAESs. As noted earlier the faculty, graduate students, and staff conducting agricultural biotechnology research in the land grant universities grew substantially between 1982 and 1988. Full-time equivalent (FTE) scientists in biotechnology have more than doubled from 273 in 1982 to 682 in 1988. Similarly, graduate students and staff working in this area increased from 418 to 987, and from 472 to 1131, respectively. The percentage of faculty in these institutions doing research in biotechnology rose from 5 percent in 1982 to 11 percent in 1988. Paralleling this growth was the increase in total funds for biotechnology which grew from 4 percent of total research funds in 1982 to 11 percent in 1988. Projections for additional FTEs devoted to agricultural biotechnology, which have been conservative in the past, indicate a further 20 percent increase in the next two years (NASULGC, 1989). Since overall faculty FTEs in SAESs grew by only 65 to a total of 6128 during this period, it is clear that the increases in biotechnology positions had to occur at the expense of other research programs. Interviews with SAES directors by the authors make clear that many of these positions were obtained by reducing the scope of conventional breeding programs. As plant breeders retire, many have been and will continue to be replaced by molecular biologists. The number of FTE plant breeders dropped precipitously between 1982 and 1986, from 415 to 298. Some of the largest programs in states such as California, New York, Illinois, and Michigan experienced losses of over 50 percent in their plant breeding community. There has been a slight increase in the number of plant breeders since 1986 but the total still remains considerably below the level of the early 1980s (NASUGLC, 1989). As noted in chapter 3, a similar process appears to be occurring within the ARS.

These developments raise a number of troubling questions. While many of the plant breeding positions have been characterized as no longer necessary, bringing the products of biotechnology to full fruition is likely to require more rather than less breeding research. Furthermore, the training of future breeders will be problematic if their numbers in educational institutions continue to decline. While there is a growing demand for plant breeders in the private sector (Kalton, 1984), the private sector does not want to train them. Moreover, plant breeding cannot be learned solely from books; one must have practical experience, gained from participation in breeding programs. Without public sector varietal development, students will be unlikely to obtain that experience.

Furthermore, it is unclear who will do the breeding for the minor crops. The problem of minor crops is very similar to that of orphan drugs. The market is either not big enough to be profitable or is not sufficiently profitable to attract private investment. As a result, the private sector invests most heavily in major crops and pressures the public sector to focus on these crops as well through grants, contracts, consulting, and even legislative pressure. The public sector also focuses on major crops as a result of state and national funding practices. Furthermore, since public funds are available for biotechnology programs for only a few crops, research is likely to follow the interests of the most powerful interest groups which tend to represent the major crops rather than those interested in increasing either the role of minor crops or the number of food crops available for human use. However, some private sector breeders are aware of the problem and want the public sector to continue breeding minor crops (Kalton, 1984).

A significant increase in the concentration of scientific talent at a small number of public and private institutions is likely. Every state can afford and has had a conventional plant breeding program. Every state cannot afford and will not be able to have a comprehensive plant biotechnology program. Instrumentation costs are particularly expensive. A survey of 41 land grant institutions in 1988 indicated that start-up funds per faculty member in biotechnology were as high as $250,000, with an average cost almost twice that of agricultural faculty members in other areas (NASULGC, 1989). Minimum funding levels of $100,000 per year per grant are necessary, according to SAES directors, for adequate funding of biotechnology research. Starting salaries are much higher for scientists trained in the new biotechnologies than for those trained in conventional breeding programs, due to their relative scarcity and their higher demand in the private sector. These factors have led to concentration of scientific talent in a few states. As of 1988, approximately 50 percent of the FTEs devoted to agricultural biotechnology in SAESs were located in just 10 institutions (NASULGC, 1989). The previously decentralized character of agricultural research insured that needs of a regional or local nature would receive attention by scientists trained in the latest technologies and methods. The trend toward increased concentration of research funds and faculty may inhibit this from occurring. In the long run, however, the situation should change considerably. First, the demand for new biotechnologists will decline somewhat as some venture capital firms go bankrupt. Second, the supply of biotechnologists will begin to increase as some universities are able to mount needed training programs. Third, as there are no restrictions on entry into biotechnology, as there are on medicine and plumbing, the salaries of new entrants will decline. However, instrumentation costs should decline faster than the salaries of scientists, allowing replacement of some scientists with machines and/or lower paid technicians. Nevertheless, it is unlikely that the concentration of scientific talent will be reversed.

The University–Industrial Complex

The scientific and commercial potential of biotechnology and decreasing federal monies for research, have led to increasing university–industry ties (Kenney, 1986). The private sector provides 13 percent of all funds expended for biotechnological research in the SAESs, while, in general, the national average for all university research is 3–4 percent (NASULGC, 1983:12). In addition to higher funding levels, the structure of private sector grants has changed. Previously, most grants were fairly small, went to individual scientists, and came with few overt strings attached. For example, a seed company might give $10,000 to an SAES for vegetable breeding with the idea that any varieties developed and any knowledge gained would be beneficial to the industry as a whole. Now, however, much larger grants are often given to a department or institution but come with strings attached. The research, even though it is often of a basic nature, is viewed as proprietary. Any patents for products or procedures developed from such research are, at the least, shared with the granting company. The potential profit to be generated from such patents is enormous. Many SAESs are spending a great deal of time and money determining how to establish guidelines for a university–industry research contract. In general, SAESs view royalties and licensing fees from patents and the legal protection of new plant varieties as a potential source of fiscal support for research (NASULGC, 1983).

The specter of the money to be made has inhibited the flow of information among these scientists. This is particularly true for those with private sector grants. Scientists must delay public discussion of such work or its results until it has been reviewed by the grantor. Even some scientists with public funding feel inhibited in talking about their work or their research ideas, believing that some private company with the money, equipment, and time might pick up on their ideas and do the necessary experimental work before they can. As one public sector breeder said: "I usually attend the Tomato Breeders Roundtable which started in the early 1960s, met yearly, and included about 100–125 scientists. About four years ago the roundtable began meeting every other year because there were fewer people working in the public sector and with the switch to greater private sector research, there was greater secrecy" (interview). In one study, 25 percent of industrially supported biotechnology faculty reported that they had conducted research that belonged to the firm and could not be published without prior consent. Moreover, 40 percent of faculty with industrial support reported that their collaboration resulted in unreasonable delays in publishing (Blumenthal et al., 1986a, 1986b). The harm done by such dampening of scientific communication is not known, but many scientists in both the public and the private sector view increasing secrecy as very harmful to the scientific enterprise. Indeed, most maintain that the free flow of new information is and should remain an essential feature of university research.

The investment of venture capital in the SAESs, almost always tied to exclusive release of technology via patent rights, is viewed as an even more disturbing development than decreased communication by many scientists. In our interviews, both public and private sector scientists stressed the detrimental effects of granting private patents for work done in the public sector. They wondered whether it represented a conflict of interest and stressed that work done in the public sector should be available to all taxpayers. In addition, many thought it might unduly influence public sector research toward individual private sector goals. As one scientist said: "This establishes a dangerous precedent; it diverts researchers from unbiased basic research of benefit to the general public to biased research oriented to specific money making objectives. It also uses tax funded money and facilities for the benefit of the exclusive grantors" (interview). As a consequence society may pay twice: once for the research and again for its benefits and products.

Such strong sentiments were not restricted to the scientists themselves. A number of research directors for seed companies stressed that private sector grants should come with "no strings attached". In addition, in an ASTA survey of its membership, a large majority of the seed companies that responded were strongly against the use of venture capital for SAES research when it was tied to exclusive release (ASTA, 1984).

Yet another consequence of closer public ties with industry may be the reduction in long-term research which traditionally has been emphasized by the public sector, in favor of research with an immediate monetary payoff. Proprietary goals have confined private funding of most projects to one year or less. In contrast, publically funded projects are generally of a longer-term nature. Dependence on private sector funds is likely to change not only the duration but also the stability of funding.

Finally, distinctions between the public and private sectors are becoming blurred by a number of recent developments. Faculty members have formed their own biotechnology firms and public universities have established for-profit corporations to develop and market innovations arising from research. Moreover, 33 states have established centers or programs devoted to biotechnology, most of which require university–industry collaboration as a condition for receipt of research funds. At the federal level, the Federal Technology Transfer Act of 1986 (Public Law 99–5012, 100 Stat. 1785) and Executive Order 12591 (10 April 1987) (3 CFR, 1987 Comp., p. 220) require government research agencies engaged in extensive biotechnology work to establish close collaboration with private companies. As the senior advisor to the Deputy Director for Intramural Research at NIH recently stated, "The danger is in excess and in the power of money to affect how we think and what we do" (Booth, 1989:20).

Dilemmas for Farmers

The new biotechnologies are also likely to have rather dramatic effects on farmers. As the production of finished seed is moved more and more to the private seed industry, product differentiation of the type found with consumer goods is likely to invade the farm sector. Thus, farmers are likely to be faced with a bewildering array of seed varieties. Already, some experiment stations release "sister varieties" to seed companies on an exclusive basis. These varieties are genetically only marginally different from those released under the experiment station name. Companies have been formed that specialize in the identification of varieties through electrophoresis.

In addition, advertising is likely to play a much greater role in seed sales than it has in the past since newly patented varieties may be agronomically indistinguishable from others already on the market. Of particular concern is that it may be difficult for farmers to distinguish between product differentiation and significant varietal differences.

Perhaps more seriously, farmers are likely to be gradually eased out of their traditional roles as the primary clients for plant breeding research. Farmers already are and will continue to be replaced by seed companies and the chemical companies that run them as well as by processors. These large companies tend to have the capital needed for biotechnology research; they can thus influence the kind of research done. For example, consider the increasing processor interest in the breakthroughs and proprietary advantages made possible by the new biotechnologies. Most biotechnology companies working with tomatoes are trying to increase total solids, which greatly increase case yields, that is, the amount of canned product per unit weight of harvested tomatoes but not necessarily crop yield. An increase in total solids of 1 percent is worth an estimated $75 million annually (Davenport, 1981). Since processors either pay no, or a very small, bonus to growers for increased-solids tomatoes, the processors gain the vast majority of benefits from higher-solids tomatoes. Furthermore, canners will only buy those tomato varieties that appear on their processing industry's acceptable list, thereby heavily influencing which varieties farmer grow.

Surprisingly few scientists or administrators appear aware of the potential for conflict between the interests of farmers and those of agribusiness. This potential for farmer–industry conflict was manifested in Senate testimony given on behalf of the American Farm Bureau Federation. The Farm Bureau argued that public research should be focused on reducing the cost of inputs, particularly chemicals (Hawley, 1984). Conceivably, the new biotechnologies *could* be used to do just that. All the evidence, however, suggests that just the reverse is happening. And, it is clearly advantageous to the chemical companies to sell more rather than less chemicals.

Perhaps a better example is the controversy surrounding the use of bovine growth hormone (bGH), an engineered hormone administered to dairy cows so as to increase milk production. The four companies vying for dominance in the market, American Cyanimid, Elanco (a subsidiary of Eli Lilly), Monsanto, and Upjohn, maintain that use will increase milk production while posing no safety threat. Many farmers contend that it will only benefit larger dairies and the chemical companies while driving family farms out of business. In several states these farmers have lobbied legislatures to prohibit its use or to require special labeling. Some processors have also expressed concern over its use (Sun, 1989). Similar concerns have been raised in Britain.

The new biotechnologies will also result in fuller integration of crop production into the larger industrial system. Seed–chemical–machinery packages are certainly likely. The little control that farmers still have over the production process is likely to be eroded still further. At the same time, some large chemical and pharmaceutical companies may develop uniform input packages that will be sold only directly through their own distributors. In so doing they would bypass farmer cooperatives and small input supply firms. These changes would undermine the access of cooperatives to inputs at reasonable prices and their very capacity to provide an alternative force in the marketplace. These standardized input packages would lead to greater uniformity of crops in the field (Lacy and Busch, 1988). This could potentially increase both food system vulnerability to climatic variation and input distribution breakdowns due to natural disasters or civil disturbances.

Restructuring World Markets

As we suggested at length in chapter 6, the growing use of biotechnology will very likely have severe destabilizing consequences for agricultural markets worldwide. Rapidly changed geographical limits to production or *in vitro* production of crops formerly produced only in Third World nations may well destroy their often fragile export economies. Lack of access to varieties produced or improved through biotechnology as a result of patents, secrecy, and cost may significantly disadvantage Third World nations in international markets. Similar, although less traumatic, changes can be expected in some parts of the First and Second Worlds. Even when the technologies or their products are accessible, the restructuring of agricultural production as a result of the increasingly rapid introduction of varieties and hybrids produced with the tools of biotechnology will surely more significantly affect the economies of Third World nations more severely than they affect those of the developed nations. Moreover, those nations with already industrialized production systems will be able to adapt to further industrialization more rapidly than those nations with nonindustrial or weak industrial production systems.

Other features of the world market may be affected as well. The investments that the United States, Japan, Germany, and other nations have made in biotechnology (see chapter 1) reflect the national industrial policies of these nations. As each seeks to be the world leader in biotechnology, other industries, perhaps even certain parts of the agribusiness sector, are likely to suffer from declining demand for their products. Even among Third World nations there are likely to be winners and losers. Some Third World nations have the scientific and industrial capacity to act as intermediaries between the First World and the very poorest nations and are already positioning themselves to take advantage of that opportunity. While it is far too early to make specific forecasts, there is little doubt that much is at stake for both winners and losers.

In summary, the new biotechnologies may well make possible substantial increases in the production of even heavily researched crops. These impacts will only be realized with concomitant changes in the organization of public agricultural research. Disciplinary relationships are also being rethought. Linkages between government, university, and industrial research are being rapidly reformulated. The danger exists that the private sector may more fully influence the direction of public sector research than it ever has in the past. This work may potentially cause further concentration in farm structure and lead to increased industrialization of agriculture. Recent work has shown that public sector research has benefited larger producers at the expense of smaller ones (Busch et al., 1984). Biotechnological research may exacerbate this trend even further, with the highly monopolized input and output sectors of the agribusiness community capturing the bulk of the benefits.

Reductionism

In his book *Algeny*, Jeremy Rifkin (1983) argues that the new biotechnologies are only a metaphor for an increasingly reductionistic, deterministic view of the world. The idea that living beings are reducible to, or at least to be explained in terms of, their molecular and ultimately their physical components is a longstanding view in the philosophy of science. Rifkin contends that this not only undermines the "mystery of life", but it has serious political and moral consequences as well. As we have shown throughout this study, the potential for molecular genetics and its agricultural applications to replace both scientifically and institutionally more holistic plant breeding techniques is great. How far the reductionist program in the plant sciences extends beyond current plant improvement strategies is, however, an open question. Furthermore, reductionistic, deterministic methodologies *have* had many significant beneficial consequences as, for example, in medicine. The expectation and the hope of reductionistic agricultural scientists is, undoubtedly, that the same sorts of improvements in the quality of human life will result from the application of molecular genetics to plant and animal life as well.

In any event, it is at least a possibility that the agricultural sciences, and in particular, plant improvement, will come to be more and more reductionistic, deterministic, even mechanistic. Many ecologists and plant breeders voiced this concern in our interviews. Even with the growth of knowledge, the power of the increased ability to manipulate genes, and the promise of "new and improved" crops, the result may be that we simply trade one form of knowledge for another. The implication is that traditional plant breeding is not only potentially important but, in fact, may reflect certain arts and skills that a reductionist view of plants – and the world – can never capture or accommodate.

While there are undoubtedly many other present and future impacts of biotechnology in agricultural research to which we might draw attention, our aim is not to present an exhaustive catalogue of highly probable events, possible events, and "what ifs". Rather, we submit that historical precedents, present trends, and future directions of agriculture and agricultural research are neither fixed nor inevitable, but are the result of individual and collective decisions that scientists, administrators, business people, politicians, farmers, and ordinary citizens are currently making and will continue to make in the near future. Decisions are affected by a variety of motivations and conditions. Ethical considerations are surely among them. Consequently, a consideration of ethics and values is critical.

Personal Morals and Scientific Ethics

On the surface, it may seem to make little or no sense to discuss the ethics of biotechnology in agricultural science. As we noted, biotechnology is a set of "tools", and hence, whatever impacts biotechnology has are the impacts of a broader practice within the agricultural research community on science, institutions, society, and the environment. Ethical decisions regarding integrity, honesty, self-restraint, love, and the like play little part in these impacts. Or perhaps, they only play a part to the extent that, for example, a particular research administrator deliberately misleads a scientist, fellow administrator, legislator, or interviewer concerning some action taken as a part of his or her role as administrator, perhaps overestimating the positive benefits of a certain program yet concealing known negative aspects. Taken in isolation, actions at the personal level, however unethical or immoral one might judge them to be when contrasted with, for example, the Ten Commandments or the Golden Rule, present no serious issue vis-à-vis the impacts of biotechnology on science, agriculture, or society at large. We can expect no greater commitment to moral values such as honesty or integrity on the part of individuals in the scientific research establishment than we can expect from those outside it.

There is, however, at least one area, conflict of interest, where the question of personal ethics is at least somewhat important in assessing the new agricultural biotechnologies. The question of conflict of interest has been raised in a legal context, and our interviews have suggested this as an important concern among scientists and administrators across the country. There is, perhaps, no more important issue of direct concern to scientists than the existence of conflicts of interest in the practice of science: indeed, the fundamental commitment to scientific honesty and candor would condemn such conflicts in the extreme.

What *is* a conflict of interest? Whatever interests we might have – in survival, security, knowledge, professional advancement, family, and so on – conflicts inevitably arise. Some interest may have to be sacrificed, or at least downplayed, in order to act on or satisfy another. These are inevitable tradeoffs. However, it is clear that there are some interests which, to the extent that they conflict, raise serious moral or ethical questions. How shall we determine when a conflicting interest situation becomes a conflict-of-interest situation?

Margolis (1978) offers some help on this dilemma, maintaining that to say a conflict of interest exists is to say the community or society has established that there are some interests that cannot legitimately be pursued simultaneously. In other words, we, in society, have determined that it is unethical for a politician to use his or her political office to gain or share information, otherwise not available, to reap a profit. What is usually at issue is that there is some well-defined personal obligation by virtue of a role, commitment, contract, or the like. The unethical nature of the conflict-of-interest situation arises from an action that is inconsistent with obligations arising from that role. The very fact that a conflict of interest might occur imposes an obligation on a person to avoid acting on the illegitimate interest.

Clearly, society's appraisal ultimately defines where and when conflict-of-interest situations can or may arise. Some of these conflicts of interest can be prohibited by law or are the basis for legal action. Moreover, what counts as a conflict of interest can vary from society to society. However, there are clear, or relatively clear, conflicts of interest identified within our society. Violation of a public trust in favor of self-interest is one. Carrying proprietary information from a previous employer to a present employer may present the possibility of another.

One important aspect of Margolis's suggestion is that the existence of conflicts of interest may indicate there are conflicting interests in society. Thus, in a fairly homogeneous society, where everyone shares the same goals, values, religious beliefs, economic benefits and burdens, and where everyone at least tries to act on this community sentiment, no conflicts of interest will arise. That is, it would not be considered illegitimate to pursue different interests, since they do not conflict. However, in our society, where divergent values, goals, and economic interests do exist, conflict-of-interest situations inevitably arise.

The importance of personal ethics vis-á-vis the use of biotechnology, and the impacts of this use on science, research institutions, agriculture, and society, ultimately rest on this point. The question must be asked: do scientists pursue biotechnological research and development on the basis of legitimate interests to further scientific knowledge or, at least in the case of public sector scientists, for the missions of public research? Or, do they do so with the hope that discoveries (e.g., new varieties, stronger and different kinds of resistances) will ultimately produce economic rewards in the form of grants, gifts, or consultantships for themselves? Perhaps the two motivations are not always incompatible. However, it is more likely that the two motivations are incompatible, at least in so far as we, society, have made the judgment that public research is to be directed toward public benefit. Indeed, most of the scientists we interviewed voiced this commitment of mission. Thus, conflicts of interest are a real ethical concern, not solely with respect to the issue of biotechnology's impacts, but perhaps even in the practice of biotechnology. As we noted, the vast sums of money available for this kind of research from the private sector make the pursuit of two interests, illegitimate when pursued simultaneously, a constant possibility. Therefore, personal ethics are important in so far as scientists' choices concerning which goals to actively pursue affect the extent to which biotechnology and public–private collaboration have their projected consequences.

Yet, if conflict-of-interest situations imply conflicting interests in a community, then this militates against an attribution of responsibility to scientists for many negative impacts we might see. Ultimately, institutional actors are more responsible for actions or decisions taken than than any one or any group of scientists.

Social Values and Institutional Ethics

We frequently and casually attribute causal responsibility to organizations and institutions by either saying that "the Experiment Station did such and such", or, "that corporation acted in a certain way". Similarly, in saying that the university, experiment station, corporation, or the like is "to blame", we make moral responsibility claims about organizations. Yet, we are not always clear in what we mean by these attributions of responsibility to institutions. It is fairly clear that the actions of individuals are capable of being assessed, both causally and morally or ethically. However, institutions, as collections of individuals with roles, authority relations, predetermined tasks, and so on, are difficult to assess in the same way. When we speak of the impacts of biotechnology in terms of institutional actions or corporate maneuvers, the important ethical question turns on our deciding who or what exactly is causally and ethically responsible, that is, who can be praised, blamed, and held accountable for those actions.

Sociologists and political theorists have long debated the nature of institutions and institutional actions. A common thread throughout these debates and analyses is that the more complex an institution and the more specific its goals and role in society, the more precisely we can talk about the organization's actions over and above the actions of individuals within the organization, whether public or private. In other words, at a certain stage in the growth and development of an institution, it begins to take on a life of its own, so that no particular individual can be held accountable, although the collection of individuals can be held jointly accountable for the organization's actions. It is in law that this idea has its clearest expression: in a corporation the legal liability of the members is limited to the extent of their contributions. The corporation exists over and above the individual actions of its members and can be held accountable in its own right (French, 1979).

The "corporate entity" view sketched above presents us with the difficult task of trying to isolate responsibility for impacts and, hence, to attribute both positive and negative ethical responsibilites to institutions concerned with agricultural biotechnology. Even more difficult is the assessment of responsibilities when, as we have noted, the impacts are the results of actions by numerous institutional actors – from SAESs to corporations, legislatures, the USDA, Congress, and so on. However, an institutional ethical perspective would require that we be able to assess relative contributions to these impacts. Indeed, the task before us is to assess the legitimacy of some of the changes and trends that we have noted. Assessing the legitimacy of the actual and foreseeable conse-quences of the actions of this system of institutions also forces us to look more closely at the legitimacy of the actions of particular institutions: for what are the SAESs, corporations, legislatures, and other actors accountable?

Assessing Institutional Accountability

So far, we have been discussing the question of responsibility in terms of the traditional and legalistic sense of attribution of causal and moral "blame". That is, when we talk about holding an agent responsible, whether an individual or corporate agent, what we mean is that we can identify some action that has been completed, that a particular agent did it, and that the agent is to be praised, blamed, or simply indicted for having done it. Jurists refer to this sense of responsibility as "formal" responsibility, and it is the standard basis for assessing liability, espe-cially criminal liability, in law.

When we speak of scientists' responsibilities with respect to actual or potential conflict-of-interest situations, we can make use of this sense of responsibility to implicate persons in the performance of illegal or unethical actions, that is, actions which have already been completed. Similarly, if we are clear on our basis for assessing the ethical or unethical nature of corporate actions toward the university, toward scientists, toward labor, and toward the environment, we can make use of this notion of responsibility as well. As we have repeatedly noted in this volume, individuals, authorities, and corporate agents are all accountable for the impacts of biotechnology on agriculture, science, and society in the causal sense of engaging in certain actions and having certain intentions. It remains for us to question the basis for an ethical appraisal of their actions. What we maintain is that for each of the impacts we have noted in this volume, *someone* is responsible for having brought them about in this legalistic sense of responsibility. Perhaps, *everyone* is responsible for them, although as our analysis suggests, some are more responsible than others for both positive and negative impacts.

However, before we proceed to our critical assessment, we must account for another sense of "responsibility". This is the sense associated with such phrases as "the responsibilities of the SAES", or the "social responsibilities of business", or "professional responsibility". Although generally referred to as "substantive" or "material" responsibilities, a better term for these is "positive responsibilities". The idea of obligation or duty is essential. To say that something is a "professional responsibility", is to say that there is something the individual, by virtue of his or her commitment to a profession, is obligated to do, think, feel, and so on. With a parental responsibility one has a duty to take some particular action or to have some particular attitude in relation to a child. Similarly, corporate social responsibilities, in this sense, are duties or obligations that the corporate agent has toward society: duties to do no forseeable harm or to maximize some benefit or, perhaps, both. Much of the confusion concerning the responsibilities of scientists, administrators, legislators, corporate executives, and the general public with respect to the impacts of biotechnology rests with a failure to understand the proper scope of responsibility, in this sense of duties and obligations. What kinds of responsibilities are there? What limits are there on positive responsibilities? And, who has the responsibility for such things as protecting the environment, preventing unfair advantage in the marketplace, and respecting the rights of labor and the consumer?

The Public Institution

We have discussed in some detail the history of agricultural research on two crops: wheat and tomatoes. As we have noted, important scientific and institutional changes have been associated with this history. Three important questions arise in regard to these historical changes. First, given the avowed mission orientation of public sector research, are the SAESs and the ARS acting responsibly by increasing certain forms of public biotechnological research and increasing private support for it? Second, are there goals or values that public research institutions should pursue that they are not pursuing and, hence, goals (i.e., obligations) for which we should hold public research institutions responsible? Third, what are the legitimate goals of public research? These are of course difficult questions, and we can only hope to sketch an answer to them here. Nonetheless, their importance demands that they be addressed.

Our interviews suggest that scientists and administrators in both the public sector and private industry clearly feel that public sector institutions have a mission and that that mission is a uniquely public one. Simply put, that mission is to serve "the public". However, this is where a clear understanding of the public institution's mission seems to end. For example, it is well-recognized that there are tradeoffs between long- and short-term research; between basic, applied, and developmental research; between teaching, research, and service; between germplasm enhancement and finishing varieties; between hybrids and pure lines; and between all the other conflicts we have noted throughout this study. And, the goals of public sector institutions seem to vary, depending on whether we are talking about the SAESs, the universities, or the ARS, and even depending on the particular institution under review. Certainly, as in the case of private industry, the institution's survival is always to be considered a primary goal, so that responses to markets, clients, funding sources, and the like, may be conceived to be rational, responsible, and even legitimate moves both by the institution and the various decision makers (causally) responsible for these actions. However, this is precisely where the main issue lies, for the other values and goals a public institution might have beyond its own survival may be in danger. These include the advancement of long-term scientific knowledge; the education of the next generation of scientists, engineers, farmers, and citizens; the preservation of a rich stock of germplasm for future generations; and the other goals and values we have gleaned from the oral and written statements of scientists, administrators, and legislators. Indeed, there is a precarious balance of these values and goals, against survival, which forces questions to be raised about the rationality, responsibility, and legitimacy of actions emanating from the shift to biotechnology and the increased private funding of public research.

Everything in this regard hinges on the clarity of purpose and the explicitness of priorities among goals. Responsible actions, those fulfilling the responsibilities of the public sector institution's mission, require a thorough self-understanding extending beyond the particular understandings of scientists, administrators, legislators, and others directly involved in roles actualizing the institution's mission. In other words, a higher-order obligation exists for those involved, and that is to "know thyself". The private institution's goals are clear – survival, growth, and profit. Responsible action is consistent with these goals. The public institution cannot proceed or be assessed without a similar understanding of exactly what should be pursued. Each participant, within the limits of his or her role, is responsible for articulating that overarching set of goals, the fundamental values according to which rational, responsible, public sector institutions are to be initially evaluated.

However, it is the *public* nature of the SAES, the ARS, and the broader missions of the university, which begins to establish a basis for that self-understanding. Indeed, the public role of the experiment station or the ARS must be specified. It has, in fact, public *roles*, and this may be what accounts for apparently contradictory actions by some scientists or administrators in the system. These actions might only reflect those different and equally correct perceptions of the responsibilities to "the public" that are part of the definition of public sector research institutions.

The tomato harvester case in California exemplifies this point. Public sector research serves "the public". The individuals and corporations that benefited by the improved tomato harvester are part of the public; yet, so are small farmers, farmworkers, and consumers. Similarly, consider research to develop hybrid wheat or to increase tomato solids content. Both seed companies and food processors are part of the public. Each of these client groups is represented in the mission of the public sector institution; thus, for public research, development, education, and extension to proceed, there must be some priority attached to the interests, benefits received, and burdens borne by each of these publics.

Let us now return to biotechnology. We have already noted that industrialization will be enhanced, concentration augmented, scientific practice revised or at least influenced, and so on. Ultimately, what will be the effects? Systematic exclusion of large parts of the public? Further concentration of science in a few, relatively wealthy institutions? Increased control of agricultural research and production in the hands of a few multinational agrichemical concerns? The impacts of biotechnological research and development raise such questions as these which ultimately reflect back on the proper roles of public sector research, education, and extension institutions. In order to "serve agriculture", those responsibilities on which public institutions are to be evaluated must be determined by constituencies benefited or burdened by the actions of the institution, in conjunction with decision makers responsible for projects, programs, and allocations of resources and scientists. Some sort of "contractual basis" for the obligations of SAESs, the ARS, and the broader university is thus implied. The public institution must be held responsible to its public financing, its public support, and for the public consequences of its actions. In short, the institution is obligated to perform according to the legitimate expectation of the broader society. This includes not only farmers, farm laborers, and input and processing industries, but consumers, urbanites, and those in nations directly affected by the work done inside and outside of the public institution proper.

This last point highlights an even more difficult question. We need to be concerned with not only *for* what, or *for* whom, public research is to be undertaken, but also with what is to be the substance of that research. There are of course clear negative responsibilities in this regard: no research should be undertaken that is foreseen to be the cause of direct and significant harm to large segments of the public. But, within this constraint, what research should be done? We will return to this question in the last part of this chapter.

The Private Firm

Questions similar to those we posed regarding the public sector institution's responsibility to its mission and to its various publics can be raised with respect to private companies and foundations, although answers will obviously differ. Are moves by private corporations into biotechnology, including increasing alliances with universities and experiment stations, consistent with the self-defined missions of those private firms? Are there goals and values that private corporations should be pursuing or actions demanded by constituencies of corporations for which the corporations should be held responsible? And finally, what is the legitimate role of the private sector in the pursuit and advancement of science and technology? These questions are certainly no less important, nor less difficult to answer than those posed for the public sector. Moreover, our task in this study forces us to at least tentatively answer them. Indeed, the present and future shape of science and society depend on the answers.

Studies such as Shetty's (1979) identify the things that observers of private firms intuitively know to be true about corporate goals. Profitability, growth, and market share constituted the top three most-cited goals for corporate actions; these were explicit in well over two-thirds of the "statements of corporate purpose" that Shetty collected from major corporations. Certainly, as we have noted, major corporations that have moved into biotechnological research as well as smaller firms that have ventured into this arena, including seed companies, must be seen as motivated by these dominant goals. Whether or not biotechnological research will in the long run produce the kinds of profits in agrichemicals, seeds, and related products that justify large capital outlays must remain at this point an open question. However, such firms, large and small, certainly *believe* that the risk in resources committed to this research will be worth it.

Yet, this is precisely where the question of the responsibilities of corporations rest. What kinds of calculations and judgments have produced the commitments to biotechnology by DuPont, Monsanto, Occidental Petroleum, and others? Are they truly consistent with the profit-maximization motive, or dictated by other concerns, even apart from market share or corporate growth? There may be no way to answer definitively this last question from outside the corporate board-room; and yet, we might reconstruct such reasonings on the basis of the consequences of corporate actions in the biotechnology area. Among these consequences, at least from the point of view of the profit structure of a major corporation, is diversification. DuPont's purchase of a seed company, and its agreement with DNAP to develop and market genetically engineered fruits and vegetables, represent a shift in the products and services provided by the parent corporation. Resources and personnel are required to integrate the subsidiary into the overall corporate structure. On a smaller scale, a seed company or even a venture capital biotechnology firm needs new talent and internal monitoring personnel even apart from capital expenditures, which we have indicated, are quite large. In both cases, the question remains: Are these indirect costs a part of directors' or managers' calculations of the investment in biotechnology?

Another consequence of the advent of biotechnology, especially with respect to public/private collaboration, concerns the amount of control that a corporation can exercise on research done outside of the corporate laboratory, in, for example, universities or subcontracted private laboratories. As experiences with other high-tech areas such as space exploration or military hardware would indicate, the potential for cost overruns, or at least, misallocations of resources, is a very real one. The farther away research activities are from direct managerial control, the greater is the likelihood that a calculable short-run return on investment will be difficult to obtain, or even ascertain. We have noted that investment in biotechnology by private firms is primarily a long-term, high-risk outlay. However, some degree of measurable achievement within budgets is essential for these investments to be continually justified with respect to the goal of profitability. Perhaps, then, profitability is not the (sole) determining goal for the investment in biotechnology.

Growth and market share are perhaps more significant goals in dictating the shape of these investments. Yet, the question of the responsible nature of these moves still remains, particularly when the dominant form of investment in research of this type is university contracts, program grants, and the like. How do long-term investments of this sort enlarge market shares and how do they contribute to corporate assets? These questions can perhaps only be answered *ex post facto*. Yet, they bear considerably on the reasonableness of the investments. Have they been part of the calculations corporate executives have made? Stockholders are certainly owed this degree of responsibility.

Earlier, we raised the specter that the private company will exercise more control over public research than it has in the past. This is important because such an increase in control is undesirable. The question to be asked here is why. This turns on the "mission" of the private sector institution and the responsibilities that institution has to consumers, labor, other institutions, and indeed, to society at large.

There is a growing literature on the social responsibilities of business, which is so extensive that only a few of these responsibilities can be pursued here. Among them are responsibilities that corporations have to their consumers to produce safe products, to avoid arbitrary discrimination against women or minorities, to avoid polluting the environment, and so on (Donaldson, 1983). However, perhaps the most important responsibility that arises in the present context is the responsibility of corporations to use their tremendous resources and power without systematically harming a significant portion of the populace. The threat of coercion or of other direct harm to individuals or other institutions is always there. The existence of such a threat and the power on which it is based provide the basis for imputing responsibilities to the corporation to use its power in a socially defensible way, that is, legitimately.

The basis for imputing social responsibilities to corporations is twofold. First, corporations have, in a sense, a contract with society (Donaldson, 1983). They are legally chartered, have rights, and can be held legally liable for actions. In this sense, they are like citizens with responsibilities. For example, citizens must not harm others. But second, and more to the point, they are *important* citizens, both in the sense that they are the lifeblood of the economy of the United States – indeed the world – and that they are capable, through even the most mundane acts, of transforming the character of a city, state, and/or nation, and of irreparably damaging individuals or the environment. Power of this sort, whether in private or public hands, imposes positive responsibilities beyond the simple "obey the law" and "mind your own business" responsibilities that are encumbent upon every citizen to follow. Corporate executives at every level attest to this.

Specifically, then, what are the responsibilities of the private sector with respect to biotechnological research and the concomitant restructuring of relationships between industry and the public sector? First and foremost is the responsibility that corporations have to act responsibly in pursuing their stated goals of profit, growth, and an increasing market share. Ours is a capitalist economy in which large, private, corporate entities provide the majority of material goods and services for society, and do so for profit. This is an historically developed fact, and we have either made the judgment that it should be so or have at least acquiesced in it. Within the parameters of that judgment, actual or tacit, the first "social responsibility of business", as Friedman (1962) argues, "is to increase the return on investment for the stockholders". This social responsibility is, however, to be acted on "within the bounds of law and common standards of morality" (Levitt, 1958).

These common standards also mean that it is socially irresponsible – in fact, immoral – for a corporation to lie, cheat, or steal, and wrong to act illegally. However, beyond these clear moral requirements there are others. Perhaps the strongest one, consistent with the nature of the capitalist system we have ratified, is for corporations not to engage in activities that systematically undermine the market system as a whole; and implicit in this is a responsibility not to alter in any negative fashion, legitimate, existing missions of public sector institutions (see Burkhardt, 1986). Not undermining the system as a whole would refer to actions such as attempts to monopolize certain markets or industries where demonstrable disutilities would result from a monopoly there. More concretely, it would also apply to corporate espionage, sabotage of competitors, and, perhaps, even to "union busting". Social responsibility with respect to public institutions is, however, in need of more explanation.

Market theorists since Adam Smith have regarded the beneficial nature of the capitalist system as based not only on its ability to guarantee an increase in the "wealth of nations", but also on its ability to permit individuals the freedom to choose their means of livelihood. However, a little noted precondition to both of these benefits is the existence of a stable, sound set of public institutions. Theorists argue over the extent of these institutions, over how many or how large they should be. Yet, the principle remains the same: public institutions, uniquely public in intent and action, are not only important, but fundamental to the existence, maintenance, and ultimately to the benefits derived from capitalist society.

Public sector institutions like the SAESs, the ARS, and universities are, as we have shown throughout this study, essential in this sense. From their inception, public institutions have provided knowledge, products, and educated people essential to the stability, growth, productivity, and profitability of private enterprise. Although private industry may complain about the extent of the incursion of regulatory agencies into private "territory", public sector research institutions have offered little such "threat". Indeed, as has been indicated, a symbiotic relationship between the university and industry is to be considered healthy, even if each is, in the process of benefiting the other, critical of the other's mission. Perhaps this is only to restate a principle equally operative in economic and political markets: competition and balance of power produce maximum benefits.

This is precisely why the foremost social responsibility that corporations have is to respect the integrity of the public sector institution. Its mission is as socially important as is the mission of the corporate institutions in an avowedly capitalist political economy. Respecting that integrity has been accomplished through many of the collaborative contractual devices currently used (NASULGC, 1985). Yet, there is cause for alarm when the fundamental mission and direction of public sector research institutions begin to be systematically altered through the actions of private industry.

While our concern thus far has been with the responsibilities of the private and public sectors to their constituencies, to society, and to themselves, one further responsibility needs to be addressed: This is the responsibility that institutions, public and private, have not only to present generations of human beings, but to the future. This is the most important issue we must confront concerning biotechnological research and the relationship between public and private sector research institutions. What responsibilities exist for the future? In short, what is the scope of our present responsibility?

The Imperative of Responsibility

We have borrowed the title of this section from a book by Hans Jonas (1984) as a way of endorsing his view that our fundamental responsibility as individuals and members of institutions is, simply, to be responsible to future humanity. What this implies is that a full awareness of the consequences of our actions, stretching as far into the future as our present vision and imagination can project, is fundamental. In the current study, we have employed a "projective analysis" (Friedland 1984) – using every conceptual device at hand in an attempt to understand as clearly as is possible where present actions, trends, and possibilities might lead. There are undoubtedly lacunae in our projections; there may be as well a certain myopia with respect to some of the impacts or possibilities associated with biotechnology in agricultural research. Even so, we conclude that there are definite areas where Jonas's imperative of responsibility entails particular responsibilities. These are the responsibilities to assess adequately (1) the future health risks associated with the use of biotechnology in agricultural research and agricultural practice; (2) the future environmental consequences of the use of biotechnology in agriculture and agricultural research; (3) the future benefits and burdens associated with the transformation of institutions; and (4) the harm to the quality of human life associated with a profound change in the nature and focus of science, resulting from increased biotechnological research and development. This last responsibility is the overarching one in our judgment: What kind of science will we leave future generations? Will it be ameliorative and humane, as those concerned would hope it would be?

Health Risks

Already, the NIH have guidelines for grading health risks associated with the use of biotechnology, in those institutions that they support. In addition, there are requirements for various levels of containment for biomaterials and laboratory equipment, safety procedures for testing and development, and so on. Obviously these are "socially responsible", although an internally generated and enforced set of guidelines and procedures encompassing the private sector would be more desirable for those institutions and individuals working in the biotechnological area. Indeed, the private sector institutions should probably generate such guidelines; without them, more governmental regulation might come into being, depending on the extent of the risk that is judged to be consistent with the public interest.

However, there is a prior question concerning these guidelines, internal or external. While some biomaterials resulting from biotechnology may present health risks, it might be argued that *agricultural* biomaterials do not and cannot exhibit these risks in the way that pharmaceuticals might. Alternatively, if they do, it is only in a very indirect way, through environmental or even economic effects on health. How, for instance, can improved varieties adversely affect health? Or, for that matter, how might a bioengineered tomato having higher soluble solids affect health? Certainly, a bioproduct might cause enough environmental harm to affect peoples' health.

Yet, this is where the responsibility for assessing future health risks comes into play. Perhaps no demonstrable risks will come to exist through the use of biotechnology. Perhaps some will. Consequently, approaches must be developed and employed for assessing and monitoring future projectable risks associated with the use of biotechnology in agriculture and agricultural research (Thompson, 1983–4). Speculation is even justifiable. If harm might result, the fundamental maxim of social responsibility would demand a halt to testing, or at least a reconsideration of safety measures. The prior responsibility is, however, to consider health-related consequences. This is incumbent upon both private and public research institutions and the individuals therein.

The Environment

Many environmental issues have arisen in the course of our analysis including: decreased genetic diversity, chemical abuse, undesired spread of modified organisms, production of toxic compounds. Some of these have obvious connections with the health risks considered above and similar points concerning responsibility can be made. However, one further consideration regarding responsibilities to the future over environmental issues is uniquely agricultural in scope.

All agricultural practice is a direct assault on the environment – on land, water, and plant and animal materials. A maxim of "do no harm" to the environment would, if taken to a logical extreme, preclude responsible agriculture, by precluding agriculture. Obviously, this is absurd. Yet, there is at least one element in this that is not so absurd, and this concerns *how* agriculture is practiced. As we have repeatedly remarked in this study, agriculture in the Western World is becoming increasingly industrialized, increasingly monocultural, increasingly dependent on external inputs in the form of machinery and chemicals. The environmental consequences of agricultural practice in these terms are increasingly becoming evident. It is incumbent upon us to consider the future impacts of such practices. More to the point, how will the use of biotechnology be linked to environmental consequences of future agricultural production? Will the role of biotechnology be positive, negative, or both?

We have noted some projectable impacts of biotechnology on agriculture and, hence, on the environment. There are probably others. What they might be is of prime concern. And hence, there exists an imperative of responsibility in this respect as well, to consider possible future environmental effects in present decisions about the use of biotechnology, and even in what kinds of biotechnological research to engage. Perhaps given our growing critical awareness of them, biotechnological research should be geared to alternative specifications of desirable characteristics in plants, and to alternative agricultural techniques and practices that might be less demanding of the environment – either in addition to, or instead of, the current trend toward research supportive of the increased industrialization of agriculture. At the very least, these other techniques and products should be considered. Moreover, when they are judged to be of more substantial benefit to the environment, health, institutions, and humans, they should be pursued.

Future Institutions

In one respect, the future is now. We have noted the rapid growth in private sector entry into the biotechnology arena, and the gradual realignment of the relationship between private industry and public sector research institutions. Inevitable questions remain about the long-term results of these new alliances. The present responsibilities of the public sector to its mission, and those of the private sector to pursue its mission and respect the mission of public institutions, are clear. Among the questions are: What do we want our future institutions and society to look like? What shape do we want them to take? What goals, values, interests, and needs do we want addressed? And, how are the benefits and burdens of agricultural research, and agriculture in general, to be distributed?

While no individual or group can provide definitive answers to these questions, we suggest that we, as a society, must not simply let historical trends "happen" on their own. In fact, they never do. Through their actions, individuals and institutions shape the present, and as a result of those actions, the stage is set for future actions. The decisions of the present set the terms and conditions for markets, negotiations, directions, and agreements for the very long term. And thus we have a responsibility to address in the most explicit terms the issue of what we want the future of agriculture and agricultural institutions to be. We have noted the inherent internal rationality of private industry pursuing profits and the reasonableness of private funding of some public research. Yet questions about the future reasonableness of such "missions" and relationships cannot be simply taken as answered. We have a fundamental responsibility to answer these questions now; but before that we must ask them.

Science

As important as the practice of science and technology is to our relationship to ourselves, our environment, and to others in our institutional arrangements, we often presume that science, like the nature it studies, should simply take its course. However, our understanding of biotechnology and the implications it has for life, agriculture, and society suggest a different view. However strongly or weakly we choose to make the claim that there are "values in science", it is nonetheless true that science affects us, and so, we are responsible for the science we practice, as well as for the knowledge, products, and processes which result from that practice. Science cannot, therefore, simply take its course; it must be addressed in terms of its probable effects.

Biotechnological research, in agriculture as in other arenas, presents us with a dilemma in this regard. Clearly, it holds promise for an improved human condition. Yet, it has also shown us that it holds promise for environmental harm, for the further industrialization of agriculture, for the intrusion of private interests too far into public sector research and education, and for the redistribution of benefits and burdens in international markets in illegitimate ways. The fault lies not with biotechnology *per se*; it is not to be dammed or praised in and of itself. Rather, responsibility lies, as with any technique, with those who move, manipulate, use, and abuse biotechnology and the processes, knowledge, and products which result. In the final analysis, the impacts of the "new biotechnologies" in the present and in the far distant future are our *present* responsibilities. Responsibilities to future generations of human beings – and, perhaps, to plant and animal species as well – demand present consideration of those impacts.

Finally, we propose one further issue of responsibility. Biotechnological research is only one aspect of an increasing trend in contemporary Western society toward a reductionist view. Thus, the question is: do we want this to be our dominant philosophical orientation when we design our future research into both external and internal nature? Although obviously somewhat speculative, this question also needs to be addressed. How shall we view the world? And what shall our science of that world be? Narrow, submolecular, instrument-dependent? Broad, global, open? Perhaps these are false dichotomies. Perhaps a variety of approaches can be integrated into a science, agriculturally oriented or otherwise, that magnifies human possibilities and the human spirit while it investigates the meaning of life in the most molecular fashion. Our responsibility to the future, as well as to the present, is to consider such a science.

8
Policy Matters

Our discussions of the potential impacts of the new biotechnologies and of responsibilities of particular actors in the agricultural research system have left open one final question: What is to be done? As we have repeatedly indicated, the nature and dimensions of the use of biotechnology in agriculture are not just individual concerns; they are matters of broader social importance. Decisions about if, when, and how to incorporate biotechnology into the agricultural research agenda are always set against the background of larger public policy issues. Public policy with respect to biotechnology may ultimately be the key to the impacts as well as the responsibilities which emerge in the practice of agriculture and agricultural science. As the title of this chapter implies, policy *matters*.

There is, to date, no comprehensive US "biotechnology policy", either with respect to agricultural applications, pharmaceutical applications, or other industrial uses. Despite a plethora of proposals for "coordinated frameworks" (Office of Science and Technology Policy ([OSTP], 1986), or "overall guidelines" (Federal Register, 1986), US policy regarding biotechnology remains embodied in a variety of plans, programs, and policies which affect research, development, commercialization, and diffusion. Moreover, different agencies (EPA, NIH, FDA, and USDA) are responsible for different aspects of biotechnology policy making and implementation functions.

In previous chapters we made cursory references to various government policies. In this chapter, we summarize in detail the various policies, programs, agencies, and institutions that constitute US "biotechnology policy". Despite the apparent incoherence of national efforts to monitor and control biotechnology, especially in agriculture, some degree of order can be imposed by categorizing efforts in terms of their generality and the extent to which they directly provide incentives, constraints and/or direction for this research agenda. Put in the context of the supply and demand framework developed in chapter 2, government policies, however contradictory they may be, define the opportunities and limits for biotechnological research. They provide rewards for certain kinds of scientific research, either directly through grants, or indirectly through tax incentives and patents. Government policies also restrict certain kinds of research through restrictions on the research process or on the sale of the products of research. In short, through government policies the broad outlines (and sometimes even the details) of science and technology are defined. It is these policies that set the stage within which science may be demanded or supplied.

In a democratic society such policies should reflect democratic values. They should help build a science that is consonant with those values. In a world in which there are an infinite number of things to be known and projects to be achieved, government policies should direct both the supply of and demand for science toward those things and projects deemed worthy of knowing and achieving.

We have seen above how government policies, client demands, and the research that scientists are able to supply, have shaped both wheat and tomatoes as well as the people and institutions involved in the research, production, processing, marketing, and even consumption of these products. We have also seen how the new biotechnologies are *at once* (1) a set of new scientific and technological facts and procedures, (2) political and economic forces on a world scale, and (3) contenders in the restructuring of our society and others around the world. Moreover, we know that many scientists are aware of and even uncomfortable with certain social, political, and economic changes that the new biotechnologies engender. This makes it all the more urgent that we understand the implications of current government policies and that we restructure those policies better to ensure that they reflect the public interest and the democratic values we hold dearly.

There are three categories that usefully delineate levels of generality and effectiveness of government policy: broader direction of economic, agricultural, and scientific efforts in general; legislative and court actions directed toward protecting intellectual property rights; and regulatory and review procedures for specific biotechnology agendas and projects. Of these, the second category has been most instrumental in affecting the course of biotechnology in plant improvement to date. However, extending an idea in a report issued by the NASULGC (1986), we will argue that a "national biotechnological impact assessment program" should be established, that would include not just *ex post facto* regulation of social, environmental or agricultural effects of biotechnology, but also an agenda that would attend to the broader socioeconomic and structural effects that might be expected. With such a comprehensive and far-reaching program, the public interest in agriculture and agricultural research could be better served.

The Larger Policy Context

As we noted in chapter 2, scientific and technical innovations in agriculture ultimately depend on the supply of and demand for those innovations. However, broader public policies regarding agriculture, the larger economy, international trade, and the science and technology agenda provide important incentives or constraints on the technology "market", even in market-based societies such as the United States. In this sense, the US government seems to explicitly encourage technological change: biotechnology's governmental (though not necessarily public) support reflects the overall thrust toward technical innovation throughout the economy. High-level research agenda setting such as undertaken by the USDA, the President's Office of Science and Technology Policy, and others only accelerates and focuses the general impetus. What is important to note is how fundamental "progress" – defined in terms of technological innovation – is to the system of governmental controls on our political and economic system.

Macroeconomic Policy

US macroeconomic policy consists of a variety of strategies and procedures designed to maintain economic growth and stabilize markets and at the same time keep inflation and unemployment at acceptable levels. Among these strategies are individual and corporate income tax policy, controls on the supply of money and interest rates, and direct government involvement in imports and exports. Although it is beyond the scope of this work to detail the mechanics of the economic processes that these and other policies are intended to influence, we need to note how economic policy affects the context in which farmers and others involved in agriculture make decisions about techniques or technologies.

Two of the more important facets of macroeconomic policy that affect agriculture are monetary and credit policy, and foreign trade. In broad terms, monetary policy consists of strategies designed to regulate the supply of money available to private banks for investment. "Tight" money increases interest rates, slows investment, and keeps inflation low; "loose" money decreases interest rates, increases investments, but has the potential to cause inflation.

The relationship between monetary policy and technological innovation, in general, is straightforward. Lower interest rates encourage new

business formation and the expansion of businesses already in operation. The development of new technologies or innovations in existing
ones, is, on the one hand, one way for new firms to break into an
expanding market. On the other hand, innovations are strategies which
firms can use to try to maintain or increase their share of the market.
Easy credit and/or low interest rates encourage consumers of technology
to adopt innovations; similarly they encourage producers of technology
to produce innovations.

Many observers of the US economy and especially the agricultural
economy in the 1970s pointed to lower interest rates and a policy which
actively promoted increases in agricultural productivity as reasons why
farmers purchased more land and acquired additional agricultural
machinery thereby furthering industrialization of production. One of
the consequences was that many medium-sized farms went (and are
continuing to go) out of business. An indirect consequence was increased demand on research institutions to deliver more technologies.
Indeed, literature on the diffusion of the agricultural technology suggests that larger farms are both more inclined to use new technologies
and more likely to place explicit demands on research institutions to
provide innovations. Furthermore, the prospects for continued economic prosperity for these larger farms, at least through the early Reagan
administration years of lower interest rates and fairly good markets for
agricultural commodities, led research institutions in both the public
and private sectors to expect a growing market for innovations that
researchers produced. Despite this, research administrators are unlikely
to explicitly identify monetary policy as an underlying reason for the
success of any technical change. Nevertheless, one explanation of why
both public and private research institutions embarked on what might
appear to be a somewhat risky research and development agenda is that
economic stability seemed reasonably assured and markets for new
technologies expanded fairly rapidly. Although agricultural biotechnology was only a promise in the 1970s and early 1980s, its potential
return was seen as substantial.

Despite the financial crisis that hit US agriculture from the late 1970s
through the early 1980s, there is still considerable incentive for firms to
remain involved in agricultural applications of biotechnology. The macroeconomic goal of increased total investment and production has been
altered somewhat and an emphasis has been placed on increased productivity – more output from the same or less input. Agricultural
biotechnological innovations like those mentioned for wheat and tomatoes are precisely the sorts of innovations mandated here: such improvements have the potential to allow more output or the same output with
fewer farmers on fewer acres. Particularly when land prices are high
and the cost of machinery and chemicals is increasing, "new and im-

proved" varieties and hybrids are attractive ways of increasing productivity. Of course, improved yields through improved varieties has long been one productivity-increasing strategy. However, the prospects of ever more rapidly produced new varieties only contributes to the "promises" of biotechnology and the demand among farm producers for these varieties.

The foreign trade aspect of macroeconomic policy contributed to this milieu as well. In the 1970s, the United States fell behind in both international agricultural and technological markets and the foreign trade deficit reached $400 billion. Given the lack of US competitiveness in other industries, especially in certain high-technology areas, agricultural biotechnology – and other applications of biotechnology as well – has come to be looked upon by politicians and academics as one way of restoring America's dominance in the world marketplace (see Vasil, 1987). Not only can biotechnologically improved commodities be marketed, but biotechnological processes can be marketed as well. Although the biotechnology industry has not been identified as a preferred industry in a US national industrial policy, it has gained many public benefits because of its potential role in an American resurgence in world trade. This role is likely to expand in the future.

Farm Commodity Policy

Farm commodity policy is part of broader macroeconomic policy, especially in so far as commodity programs such as subsidies, production limits, acreage restrictions and the like are designed to stabilize a somewhat volatile sector of the national economy. However, farm commodity policy has contributed in its own special way to the high technology thrust of the agricultural economy and sets the stage for the biotechnology "revolution". In principle, the natural or biological exigencies of farm production (i.e., droughts, blights, deluges, early frosts), as well as the importance for national security of a stable and affordable supply of food and fiber, traditionally dictated a set of strategies that sought to keep full-time farmers in business. The government, through USDA and various subagencies was the main contributor of this policy and guaranteed what agricultural markets could not: reliable income for farmers. Some of the strategies included providing loans against future crop revenues for working capital, insuring crops against natural disasters, and subsidizing market prices or restricting output to guarantee "fair" market prices. Since the 1930s, when these policies began, the government has continued to play an active role in the agricultural economy, although the emphasis has shifted from keeping farmers in business to making sure that farm commodities are available in the marketplace at reasonable prices for consumers. The effect of this subtle change is that farms that are better able to manage their own financial resources, to produce at lower costs, and to guarantee (within the limitations imposed by nature) a steady supply of produce to processors or distributors have become the major beneficiaries of farm policy. Although there continues to be some concern among small-farm advocates in Congress that small, family-sized farms remain in business, most observers now recognize that larger farms in fact reap most of the benefits of US farm commodity policy (Gardner, 1981).

How the exceedingly complex commodity policy system contributes to industrialization and concentration in general is beyond our scope here (however, see Strange, 1988). Our point is that the implications of current farm policy for the biotechnological "revolution" in agriculture are clear. Farm policy encourages farm producers to produce as cost effectively as possible; it also encourages farmers to be as flexible as possible in responding to market changes. This means that farms must be able to more effectively control production. As we showed earlier, biotechnologically improved varieties or hybrids are likely to give farmers more factory-like control over their farming process. Although farm commodity policy cannot be said to cause the increase in farm size or dependence on machines, chemicals, or new plant varieties in any direct way, it nevertheless favors those farm operations that already benefit from economies of scale (Penn, 1979).

Few would accuse the authors of the farm bills of intentionally discriminating against smaller farmers or intentionally promoting more industrialization in agriculture. In fact, the justifications for farm policy frequently point to the plight of marginal but operating family farms. However, it is clearly recognized by most farm policy analysts that the consequences of farm policy may hinder the competitiveness of smaller farms. These same consequences will promote biotechnological innovations in farming.

Science and Technology Policy

It is against the background of macroeconomic and farm commodity policies that the implications of science and technology policy for biotechnology in agriculture become clear. Nearly all developed and some less-developed nations maintain high-level governmental agencies or advisory boards to deal with scientific development. These institutions generally promote scientific advancements in areas critical to the national interest, for example, those with implications for defense, health, or energy. In the United States, at the federal level, several agencies and advisory boards, representing those branches of government and cabinet-level departments where science is of direct concern, set and direct science policy. Moreover, even within a department or governmental branch a number of separate "science bodies" may exist. For example, not only do the US Senate and House of Representatives have standing committees on science and technology, but there are also separate subcommittees on military technology, health-related technology, and science and energy. The president's office contains the Office of Science and Technology Policy and consults various other groups such as the National Academy of Sciences and National Science Foundation on a regular basis. Also, the Departments of Health and Human Services, Defense, and Agriculture each have at least one science policy office. And, most major scientific professional organizations routinely offer testimony and give consultations when legislation is pending which will affect science or technology development.

The crucial feature of science policy, whatever its institutional source, is that scientific advance, indeed, progress, is explicitly mandated by it. Although particular sciences, especially physical and biological sciences, receive greater endorsement both in public pronouncements and in public resources made available, all sciences including the social sciences are promoted by the operation of science policy. Emphasis frequently shifts back and forth among disciplines, projects and directions. For example, at the highest levels of government, space-related science might be targeted one year, medical or biological science aimed at the cure of cancer another. Nevertheless, the government plays an active role in contributing to the increase in knowledge in all scientific disciplines, through a wide variety of projects, directions, and research orientations.

However, another significant feature of government-promoted and government-sponsored science is, that while so-called basic science has been targeted as the major direction for US research, most science policy has a strong technology development component as well. For example, the Technology Transfer Act of 1986 mandates public–private cooperation. This emphasis on technology is especially relevant to agricultural science, given the government's commitment to increased total production and productivity in the food system. There is no hard and fast distinction between basic science and technology development: an engineering success may have clear implications for basic physical theory, while discoveries in molecular biology may have profound agricultural applications. However, the point here is that US science policy favors applications. National interest and national security are maintained and strengthened not only by growth of knowledge or information, but also by the development of medical techniques, military hardware, communications technologies, and agricultural production strategies.

Agricultural research policy – as established by USDA, state departments of agriculture, and at the local level by colleges of agriculture – is a prime example of the general thrust of science policy oriented toward the development of useful techniques and technologies. In recent years, USDA has emphasized basic science and has allocated a significant amount of resources for biotechnology research in agriculture, thereby reinforcing its importance. However, the ultimate criterion by which even basic scientific discoveries are judged is still whether or not they will result in returns to agricultural production. In this regard, biotechnology research, while in many respects a basic scientific enterprise, is evaluated in much the same way as other technology-development projects such as those in agricultural engineering.

It is well known that macroeconomic policy, farm commodity policy, and general science policy significantly affect both the structure of agriculture and overall technology development. The key issue is that these larger policies are seldom included as part of a technology assessment program. It is generally assumed that the economic system will progress in beneficial ways according to market processes under the watchful eye of the government. It is further assumed that agricultural productivity and food security will result from the successful introduction of more science into agriculture. And it is assumed that scientific progress mandated by the market and governments will generate technologies that contribute to prosperity and abundance. However, as we have suggested above, the adequate assessment of any technology also demands the assessment of the institutional structures that have led to its creation, promotion, adoption, and development in the first place. Resolving the issues associated with the new biotechnologies in plant improvement depends on our answering more basic questions concerning the nature and direction of many of the fundamental policy components in our political economy. In particular, is growth, through technological innovation, ultimately in the public interest? Is increasing agricultural production and/or productivity through increased applications of technology the right way to proceed? Is promoting science and technology development, without regard to social effects, justifiable? Is biotechnology just another tool? Clear answers to these questions should govern our future economic, agricultural, and science policy.

The Legislative and Judicial Context

A "biotechnology as technology" or even "biotechnology as commodity" orientation is clearly manifest in a number of Congressional actions and judicial decisions which have given both public and private research efforts further incentive to develop agricultural applications of biotechnology. Public funding of biotechnology was discussed at length in chapter 1. In addition, both congressional and judicial actions have focused on the intellectual property rights aspects of the new biotech-nologies. In particular, the PVPA of 1970 and a series of Supreme Court and Patent Office decisions with respect to the patentability of "novel life forms" have increased – and in some cases caused – the develop-ment of new plant varieties. In both the public and private sectors, the potentially high revenues from being the first to develop, register or patent, and market a new plant variety with novel attributes has under-standably produced a race among scientists and firms. As we indicated in chapter 1, the influx of corporate and venture capital into plant improvement efforts has been enormous. Much of this influx is the direct result of the PVPA. Asexually produced plants have been patent-able since the passage of the Plant Patent Act (PPA), an amendment to the Patent Act of 1930. However, the speed of innovation in plant improvement was not significantly affected, nor were excessive profits (monopoly rents) reaped by those who did patent new plant varieties (Stallman, 1986). This is in part because the technology for asexual propagation was the traditional grafting of fruit trees and berry vines (e.g., grapes and kiwi), enterprises limited to a few species. Moreover, because the patented product was widely available for use as a graft itself, phenotypically very close although legally "distinct" plants could be marketed within a very short time by a variety of breeders. Neither the technology for identifying the precise "distinctiveness" of plants nor much incentive for enforcing the patents were available. As a result, the PPA had little impact on plant breeding or on the market for plant varieties (Stallman and Schmid, 1987).

The PVPA extended property rights to breeders (commonly known as plant breeders' rights) for all sexually propagated plants, except F_1 hybrids. The subsequent 1980 Supreme Court decision *Diamond v. Chakrabarty* (447 US 303) allowed the patenting of "novel life forms'. Based on that decision, in *Ex parte Hibberd* (1985) the US Board of Patent Appeals extended industrial patents to all plant varieties and hybrids. Given the tremendous time and costs associated with tradition-al breeding, whether of open-pollinated varieties or hybrids, these deci-sions may not have had any impact on plant improvement efforts either, had it not been for the success of recombinant DNA techniques. As we have already noted, this technology permitted two things: first, varieties and hybrids could be produced more rapidly; and second, plants that were genetically very different but phenotypically very similar could be distinguished through the use of sophisticated mapping technology. The owner of a patent can now prove whether a variety developed by others, either through traditional breeding or rDNA technology, in-fringes his or her patent rights.

These judicial and legislative actions have created considerable excite-ment among plant breeders and plant molecular geneticists in both the public and private sectors. Nearly every scientist and administrator we interviewed said that these events could significantly affect plant improvement programs: the prospect of introducing an economically successful patented variety on the market could generate considerable revenue for a scientist, university, or corporation. There is little wonder that scientists, including traditional breeders, are inclined to reach for any scientific tool that might speed up the breeding process – particularly biotechnology.

However, the immediate short-term consequences of the PPA, the PVPA, and the *Chakrabarty* and *Hibberd* decisions on plant breeding, and even on the structure of agricultural research in general, are only part of the picture concerning property rights in plants. Equally if not more important are the broader, long-term and, in some cases, worldwide implications of plant patents. One of the major concerns expressed by scientists is that intellectual property rights, which are intended to provide an incentive to research a particular area, may actually inhibit the flow of information among scientists, information that is necessary for further research. Private firms are justifiably reluctant to share the germplasm they have improved, but they will probably become increasingly reluctant to share information about other materials they hold. Indeed, information that is shared with other scientists, whether in the public or private sector, is in many cases limited. Public sector scientists with corporate grants may even be unwilling to share research assistants for fear that trade secrets could become jeopardized. These issues are of course not limited to intellectual property rights in plants; nevertheless, they are apparent in this connection.

A more far-reaching issue is whether or not individuals or institutions in the United States holding industrial patents on varieties will make those varieties available, either freely or for sale, to scientists and agriculturalists in developing nations. Companies holding patents on elite breeders' lines might force relatively impoverished Third World scientists to purchase germplasm for their scientific efforts at artificially inflated prices. Even more significant would be collecting patent royalties from nations that desperately need a particular pest- or drought-resistant variety. Of course, these and many other issues concerning property rights and patents extend beyond plants, no matter how they are developed. Nevertheless, the future potential of biotechnology, the rapid creation of new plant varieties, and the possibility of industrial patent protection force us to confront the issues of seed use and availability. Breeders and farmers, no matter where they live, need seed, and if it is not available they are the worse off.

The PVPA and the other patent protection mechanisms also force us to raise questions concerning the availability of germplasm for future generations. Some critics have questioned whether biotechnology will

Our
Good Taste
is a Product
of Culture

Thanks to tissue culture research we now have the capabilities to produce superior hybrids at an accelerated pace. In only three years Arco researchers have developed new cabbage and carrot varieties that have better uniformity, higher yields and improved disease resistance.

So what can growers worldwide expect from Arco Seed in the coming years? More new products and better hybrids than ever before.
At Arco, we know the future belongs to those who research it.

ARCO Seed Company
Subsidiary of AtlanticRichfieldCompany
619-352-2091

Figure 8.1 This ad promotes the new technologies as the way to produce superior hybrids for a global market.
(© ARCO. Reproduced by permission.)

increase or decrease genetic variation in cultivars. Of course, it need not do either. However, biotechnology in plant improvement, in connection with expanded property rights in plants so improved, may lead to less concern with varieties with no currently discernible value. Without potential agronomic value, there is less incentive to improve, maintain, or even sustain a variety. Yet, abandoning such varieties will both diminish our heritage and leave us vulnerable to as yet unknown plant pests and diseases. Increasing the vulnerability of future generations through decreasing the stock of variability in genetic materials may not have been the intention of the authors of the PVPA or the jurists and administrators responsible for the *Chakrabarty* and *Hibberd* decisions. Their intentions may have been quite the opposite, in fact, displaying a concern for decreasing the future's vulnerability by increasing the frequency and utility of innovations. Nevertheless, property rights in plants may disadvantage our successors.

The final issue associated with property rights concerns the extent to which private companies, whether or not they hold patents, engage in research efforts in which the larger public may have some interest, for example, transfer of genes between food crop plants and toxic microorganisms or weeds. Though not precisely an issue of intellectual property rights in the sense of patents, the concern here is that trade secrecy in the plant sciences may *de facto* permit a considerable amount of research that is, in fact, dangerous. To date, most of the public concern with biotechnology has been in areas such as pharmaceuticals and pesticides, where the potential for health or environmental harm seems greatest. This is also the area where regulation and court decisions have been most often directed. But there are other potential social consequences of biotechnology research which may be affected by the primacy of property rights. One of the implications of our discussion is that the public has an interest in health, environmental, and socioeconomic effects associated with plants. Property rights, intellectual or otherwise, frequently need to be constrained for the public interest. In some respects they already are, given regulations on what can be researched and released, but in the future we may need to constrain such property rights further.

The Regulatory/Review Process

Governmental support and encouragement of biotechnology in agriculture has been the theme of the two previous sections of the chapter. Governmental or social control is another subject which we must now consider. Despite the importance of broader economic and property rights policy with respect to biotechnology, decisions and actions in this category have been the subject of most concern, both inside and outside the agricultural scientific community. For practicing scientists, the government's role in licensing, regulating, and testing has the most direct impact on their work. Citizen groups see the regulatory role of various agencies, as well as the power of the courts in blocking licenses, field testing, and so forth, as the most significant public policy arena with respect to biotechnology. However, much if not most of the biotechnological research and development directed toward the plant sciences is only peripherally controlled by current regulatory policy and court actions. A more thorough and comprehensive agricultural biotechnology impact assessment and regulatory policy is called for, both for the sake of scientific and technological advance and to assure that the public interest in environmental, health, and safety aspects of agriculture and agricultural research is maintained.

The regulation of biotechnology, generally, falls within the jurisdiction of three agencies: the FDA, the EPA, and the USDA. In addition, guidelines exist for the use of biotechnology in federally funded health and medical research and are administered through the NIH. The NIH guidelines specify "containment levels" for degrees of risk associated with specific rDNA projects, and thus, regulate the process of doing biotechnology; otherwise, the governing of biotechnology is achieved only indirectly, through the regulation of its *products*. Even the FDA's "good manufacturing practices" assessment standard treats biotechnology as just another manufacturing process: so long as foods or feeds produced biotechnologically contain no harmful contaminants or residues, they are no more nor less "generally recognized as safe" than other processes.[1] The FDA's influence on agricultural applications of biotechnology has to date been relatively small, confined to premarket clearance tests of animal feed additives and veterinary medical devices. In addition, the NIH guidelines are for research performed by NIH or under other government grants or contracts, and are only voluntary for private industry. Therefore, they have had a minimal effect on agricultural biotechnology in the plant sciences.

Agricultural applications of biotechnology generally fall within the purview of either the EPA or the USDA or both. These agencies also have had little direct effect on plant improvement programs, except for controlling the release of genetically engineered noxious plants and plant-affecting microorganisms into the environment. Significant legislation in this regard includes the Toxic Substances Control Act (TSCA) (Pub. L. 94–469, 90 Stat. 2003), the Federal Insecticide, Fungicide and Rodenticide Act (FIFRA) (25 June 1947, ch. 61 Stat. 163), the Federal

---

[1] However, the recent decision regarding vanilla (chapter 6) suggests that the labeling requirements of the FDA may provide a partial form of regulation.

Plant Pest Act (FPPA) (Pub. L. 85–36, 71 Stat. 31), and the Federal Noxious Weed Act (FNWA) of 1974 (Pub. L. 93–629, 88 Stat. 2148). The complexities of the definitions, the testing and licensing requirements, and the jurisdictions of these acts are beyond the scope of our work here, but we will briefly comment on the kinds of impacts these laws are designed to address even though they are irrelevant to many of the biotechnological impacts on agriculture.

The TSCA gives EPA the authority to ensure the review of all new chemicals prior to their commercial manufacture by requiring a pre-manufacture notification. The law allows EPA to obtain information from producers of chemicals, with or without the producers' consent, and to prohibit the use or limit the amount of the chemical that can be produced. Alternatively, labeling may be required. On the list of "chemical substances" covered by the act are a variety of microorganisms; however, research and development using reasonable amounts of these organisms and other toxic chemicals are excluded from the notification requirement and are not subject to regulation under this statute.

The FIFRA governs the use or release of dangerous substances into the environment, commercially or in the research and development process. It has also been the statutory basis for the lawsuits, brought by Jeremy Rifkin's Foundation on Economic Trends, against agricultural researchers involved in rDNA work. The FIFRA defines a pesticide as any substance, organic or inorganic, that works to repel, destroy, or lessen the destructive effects of a pest. Plant growth regulators are also covered by this act. The EPA has the authority to register pesticides and issue licenses to manufacturers and applicators (persons licensed to apply pesticides). It also can issue an experimental use permit for research involving pesticides, depending on an environmental risk assessment carried out by EPA itself. Once EPA determines there are no significant risks to human health or the environment from field testing a pesticide, it grants a limited license. Commercialization of the product, of course, requires a full-scale registration process and relicensing for manufacture and use.

The USDA's main regulatory efforts which affect biotechnology are through the FPPA and the FNWA. The intent of these laws is to limit the use and movement of plants and plant pathogens which, while not a threat to human health or safety, nevertheless can have profound effects on agricultural production. The FPPA and FNWA require that permits be obtained for experimental or scientific use of listed plants as well as pathogenic microorganisms; in fact, the only permits that are issued are for research purposes. Since so much of genetic engineering depends on the use of pathogens and noxious relatives of crop species, these two statutes have considerable effect on the actual practice of genetic engineering in the plant sciences. Once again, however, the intent of these laws is to ensure environmental safety so that ultimately agricultural production is secured. Larger socioeconomic effects of the development or diffusion of the plants or microorganisms, if at all relevant, are only indirectly addressed.

With respect to regulating the practice of biotechnology and biotechnologically produced plants and microorganisms, there are a number of issues that are matters of public concern. One issue is the efficiency and effectiveness of the regulatory process: If there are health, environmental, or social risks associated with these new research and development strategies, let us hope that the agencies responsible for protecting the public interest do their business well. There are some questions being raised about their activities. As of early 1990, the USDA and the EPA still had not published their long-promised proposed rules and guidelines for evaluating field tests of genetically engineered organisms. Moreover, the federal Biotechnology Science Coordinating Committee, established to coordinate regulatory issues that cut across agency lines, has been active "behind the scenes trying to thwart EPA's rule making" (Mellon, in Fox, 1990:142). A second issue is the impartiality or fairness of testing, reporting, and licensing. We certainly hope that FDA, EPA and USDA do not knowingly or intentionally favor particular laboratories, firms, universities, or scientists in the degree to which their work is scrutinized for possible ill effects. A third issue is the scientific and social legitimacy of the safety standards which are applied to bioprocesses or bioproducts. The sophistication of testing and measurement procedures increases as scientific knowledge increases, and we hope that the risk analysis will be as accurate and complete as possible before these products are released into the environment.

However, as important as these issues are, our main concern is not simply that regulators do their job professionally. Our concern, instead, is that important aspects of biotechnology have been omitted from the current regulatory framework. The health, environmental, and social effects of any techniques or technology are profoundly important. While we do not propose that biotechnologically produced new plant varieties and hybrids should be prohibited from release based on demonstrable or even likely negative social consequences, some formal, institutional mechanism for inclusion of these effects in the decision-making and regulatory process is required. Otherwise, they are likely to be ignored by all but a few socially responsible scientists, administrators, or critics. They are certain to be ignored by those responsible for producing the effects.

In the (in)famous California tomato harvester case, a California state court ordered the University of California to establish an administrative procedure to ensure that federal funds were expended with primary consideration for the small family farmer. However, in 1989, that decision was reversed by the California Court of Appeal which ruled that such a procedure was not actually required by the Hatch Act or by any other legislation (*California Agrarian Action Project, Inc. v. Regents of the University of California*, 210 Cal. App. 3d 1245, 258 Cal. Rptr. 769). Nevertheless, some universities are attempting to include a social impact assessment component in their research review processes. Unfortunately, like NIH guidelines, which apply in only limited fashion to public research efforts, these institutional review boards can only guide public research efforts at institutions that voluntarily establish them. Institutional review boards should be mandated at all public institutions. Perhaps they should also be required in private industry efforts. As we discussed above, socially responsible or even professionally ethical behavior may in fact hinder public research efforts in biotechnology if those efforts are not matched by socially responsible and professionally ethical behavior in private enterprises involved in research and development. Thus, greater public participation in the total research and development system is necessary. This includes participation not only in decisions about the health and environmental dimensions of biotechnology, but in decisions about its social, economic, and political dimensions as well.

A National Biotechnological Social Assessment Program

In 1986 NASULGC, in conjunction with USDA, proposed a comprehensive formal mechanism for evaluating the issues, alternatives, and objectives associated with agricultural research in biotechnology as well as the agricultural applications of biotechnology. The stated goal is to integrate the various review procedures and criteria regarding biotechnology so that both health and environmental safety are better secured, and more effective scientific research, unhindered by "nuisance" lawsuits, can proceed. Accordingly, NASULGC and USDA proposed that agricultural applications of biotechnology be overseen by an Agricultural Biotechnology Recombinant DNA Advisory Committee (ABRAC) in conjunction with the National Biological Impact Assessment Program (NBIAP). The ABRAC advises USDA and the public on all public and environmental dimensions of biotechnology and directs USDA efforts to establish "appropriate controls with respect to the development and use of the application of biotechnology to agriculture." The NBIAP coordinates the efforts of existing organizations concerned with agricultural biotechnology and compiled a set of guidelines allowing ABRAC to determine, in advance, whether research is done under "appropriate controls". These guidelines roughly parallel those of NIH, but are extended to cover experimental release of genetically engineered organisms into the larger environment. In a sense, the guidelines themselves are summaries of EPA and FIFRA decisions regarding rDNA-produced organisms over the past few years.

Nevertheless, as aggressive as the USDA-NASULGC system might appear to be for coordinating advisory efforts in the agricultural use of biotechnology, it has two main problems. First, it lacks an enforcement mechanism: it relies on the current system of "decentralized, scientist-based responsibility for safety" (NASULGC, 1986:9) and current regulations. Thus, it is more a "clearinghouse" than a focused assessment program. And even more than the NIH guidelines, NBIAP is a voluntary system. Second, and equally important, the ABRAC remains concerned mainly, if not solely, with human safety and environmental impacts of biotechnology; social consequences, except in the most narrow sense, are not the basis for establishing "appropriate controls". As such, socioeconomic changes in agriculture due to agricultural biotechnology are effectively ignored.

Solutions to these problems are straightforward. First, a national biotechnological review and assessment program should be initiated which has strong regulatory powers. Some critics have argued that the US does not need another layer of government involved in the process of regulating biotechnology. Additional bureaucracy may slow down scientific progress and threaten the competitive position of the US biotechnology industry worldwide. However, if the preservation of public health and safety is the goal of public agencies and public policy, and this goal can be best achieved through an effective, comprehensive, and authoritative public agency, that agency should be created. If certain individuals or institutions resent this further intrusion of "the government" into their research, development, and marketing activities, so be it. They are not entitled to place the public interest at risk.

Second, the consideration and inclusion of the social, economic, and ethical dimensions of biotechnology should be at the center of the assessment and regulatory process. If research efforts, or the experimental release or the commercialization of particular products are likely to have demonstrable negative social and/or economic effects, it is in the public interest that these be duly noted. If these effects are significant, the process or product should be prohibited from use or limited in the nature and duration of its use. Given all the positive claims made for biotechnology by its proponents, this should not be too great a burden to bear.

A Public Agenda for Agricultural Research

The crucial issue is not, however, primarily a matter of assessment and *ex post facto* regulation or prohibition. Once research is performed, the results of that research are invariably used, somewhere or somehow. Of course, the strengthened national biotechnology social impact assessment institution described above would prohibit some research and seriously constrain the use of some results of research already performed. However, the more fundamental issue concerns the *ex ante* direction of our national agricultural research agenda. The public research agenda should be such that negative health, environmental, and social effects of science and technology development are prevented from occurring in the first place. This may imply that some scientific and technological possibilities, to the extent that they are linked to undesirable social and environmental possibilities, should simply be placed outside the realm of practicable science. There is always a chance that present discoveries or yet undeveloped research projects may lead to tremendous, beneficial, practical solutions to many of the problems we not only currently face but are likely to face in the future. However, pursuing or allowing the pursuit of some of these may simply not be worth the risk.

The public agricultural research agenda, articulated in a coherent national agricultural research policy, must consider both the risks and benefits that are associated, not just with particular biotechnology projects or particular products produced through biotechnology, but, with the entire agenda itself. Biotechnology, whether applied to plant improvement, animal breeding, food processing, or other areas, is only part of the larger agricultural research effort. As we have indicated many times throughout this volume, it is within the context of this larger effort that the promises and prospects of biotechnology, as well as its risks and dangers, are to be realized. The truncation of time and space in plant breeding through gene transfer and tissue culture is only representative of a desire to exert greater and greater control over the production of food and fiber; the desire to be the first to introduce a patented tomato variety or wheat hybrid derived from these new technologies is only representative of the desire to exert greater and greater control over the market. The point is that specific agricultural research projects involving biotechnology, and even entire research agendas, reflect and embody larger social commitments with respect to food, science, and even the political and economic system in which we live.

A public – and socially just – agricultural research agenda will focus on the larger question of what is really in the public interest. Evaluating this agenda is not simply a matter of administrators performing risk–benefit analyses. Risk–benefit analysis may certainly be performed, and may be helpful in understanding the nature of research, but the ultimate determinant of the legitimacy of research policy must be whether or not this policy meets the public interest in agriculture and agricultural research. What is and how we determine "the public interest" in agriculture and agricultural research are, of course, difficult issues. These questions have been and continue to be the subject of much debate. This is how it should be, because through debate public interest is both articulated and served.

The current direction of biotechnological research in the plant sciences reflects a variety of concerns, from broader economic to more narrowly scientific concerns. Yet, one theme predominates both in the stories recounted here and in our interviews held with scientists and administrators: as biotechnology became available to agricultural research, its use appeared compelling. It is as though no individual scientist or administrator, no public institution or private firm, had a choice. Even scientists with no particular training or interest in the newer techniques found themselves under some imperative to "re-tool". New institutes were created out of whole cloth – some consisting of nothing more than flashy brochures with lists of names of scientists from different disciplines. A generous conclusion is that somewhere, somebody decided to make biotechnology the key item on the national agricultural research agenda. A less generous conclusion is that everyone let biotechnology become the key item, because everyone thought everyone else wanted it to be that way. There were immediate beneficiaries, of course. Unfortunately, we believe that the process of determining national priorities, priorities which would govern both public and private research efforts, is not at present deliberate.

The process of determining a national agricultural research agenda must be deliberate and carefully considered. It must include input from a variety of individuals, both inside and outside the agricultural scientific establishment. Many scientists we interviewed expressed dismay that they were never asked about the shift in emphasis to biotechnology in plant science. Research administrators in the private sector expressed concern about how government priorities are determined. The process of determining the public interest in agriculture and agricultural research must become more public. It must also attempt to introduce clear priorities into agricultural research. Table 8.1, for example, is a list of USDA goals for agricultural research. Even at the level of generality at which these goals are articulated, there is considerable contradiction, both among the goals and within each goal. Increasing agricultural productivity may reduce the number of producers thereby undermining rural communities. Certain measures that protect crops and livestock may threaten human health and/or nutrition. The agricultural research establishment has attempted to serve many masters for a long time. Perhaps in setting the research agenda, determining priorities, and setting the course for the future, the time has come to serve one master: the public interest as a whole.

There are of course many publics and many interests among those

Table 8.1   USDA Goals of agricultural research

Insure a stable and productive agriculture for the future through wise management of natural resources.

Protect forests, crops and livestock from insects, disease, and other hazards.

Produce an adequate supply of farm and forest products at decreasing real production costs.

Expand the demand for farm and forest products by developing new and improved products and processes and enhancing product quality.

Improve efficiency in the marketing system.

Expand export markets and assist developing nations.

Protect consumer health and improve nutrition and well-being of the American people.

Assist rural Americans in improving their level of living.

Promote community improvement including developing of beauty, recreation, environment, economic opportunity, and public services.

Enhance the national capacity to develop and disseminate new knowledge and new or improved methodology for solving current problems or new problems that will arise in the future. Research in pursuit of this goal is conducted under all the research problem areas of the other goals.

Sources: Adapted from USDA, 1982c

publics. The search for the overall public interest is never easy. However, unless we, as individuals and as members of society, act on the supposition that there is a public interest which we can establish or discover, our public policies, including science and technology policy, dissolve into special protections for the more powerful among us. Duly noting the trends can help us rectify them.

Several recent attempts to establish the public interest appear noteworthy. Iowa State University's Agricultural Bioethics Committee was established "by legislative statute in 1986 to investigate, discuss, and monitor the ethical, social, economic, and environmental impacts and implications of agricultural biotechnology" (Glick, 1988:1). This committee publishes a newsletter *The Bioethics Forum*, and has sponsored a symposium and published several monographs.

Similarly, the National Agricultural Biotechnology Council, a consortium consisting of the Boyce Thompson Institute, Cornell University, Iowa State University, and the University of California at Davis, has plans for identifying the issues of national scope. The Council intends to hold annual conferences to discuss these issues.

The Keystone Center in Colorado has been holding workshops to facilitate national policy discussions on biotechnology for several years. Its objective is to encourage discussion of the scientific, regulatory, environmental, and socioeconomic issues surrounding biotechnology by lay leaders and policymakers.

The Agriculture, Food, and Human Values Society has addressed many of these issues in both national conferences and through its journal *Agriculture and Human Values*. And finally, the National Agriculture and Natural Resources Curriculum Project has addressed these issues as part of its task force on "Ethics and Public Policy in Agriculture".

These disparate efforts are clearly steps in the right direction. Yet, for all the public and private funds spent on research and development of the new biotechnologies, only a minute sum has been expended to assess their likely impacts or to develop public policy alternatives that will better serve the public good. Rectifying this situation is hardly a luxury.

Moreover, assessing the consequences of the new biotechnologies will become more difficult if we do not maintain a strong, independent public agricultural research system in which open communication of scientific information is assured. This system and public confidence in it will be hopelessly compromised if private sector collaboration takes precedence over the public good as a university goal (Lacy et al., 1988).

Furthermore, the public research system must be complemented by an extension service that provides all farmers and other citizens with the latest technical information as well as information needed to assess the larger socioeconomic, environmental, health, ethical, and other consequences of technical change. At the same time, extension needs to improve its effectiveness in engaging scientists and interested publics in a more effective dialogue concerning the directions of agricultural technical change in general, and biotechnology in particular. In short, extension must also provide an agricultural *intention* service.

A public agenda for agriculture must come to grips with the problems of the Third World. It must help developing nations to reap the benefits of biotechnology. This effort will require regulation to prevent the sale of potentially dangerous products or the implementation of unsafe testing in Third World nations by unscrupulous companies. It must also include assistance to improve both their technical research and their capacity to evaluate it in terms of their own public good. This will hardly be an easy task.

Finally, we need to strike a balance, nationally and globally, between the short-run, high-profit, proprietary biotechnologies, and the long-term, low-profit, nonproprietary ones that offer the promise of an agriculture that is environmentally sound, improves our health, and focuses on building sustainability in both our fields and our communities. To do this we must come to grips with the fact that, to paraphrase Pogo, we have met the environment and it is us.

The potential for biotechnology may be practically unlimited. As we have seen, much of that potential has yet to be realized in plant science. Scientists and their supporting institutions, whether in the public or private sector, have good cause to look to the future, to the prospects biotechnology holds for the improvement of agriculture, and to the increase in scientific knowledge. One scientist cautioned: "Look at history. We have often sailed into using products of technology before we understood the real implications. There is a direct relationship between the potential for good and evil. Science and the informed public must participate in direct checks and balances" (interview). In short, there are reasons why each of us should be concerned, and why, in the final analysis, the realization of the potential of biotechnology may require that it be directed and limited. Although few of us wish to admit it, science, agriculture, and society are human enterprises, and our second nature is frequently less than noble.

Bibliography

Abelson, P. H. 1988. World competition in biotechnology. *Science* 240: 701.

Abir-Am, P. 1982. The discourse of physical power and biological knowledge in the 1930s: A reappraisal of the Rockefeller Foundation's "policy" in molecular biology. *Social Studies of Science* 12: 341–82.

—— 1987. The Biotheoretical Gathering, transdisciplinary authority and the incipient legitimation of molecular biology in the 1930s: New perspectives on the historical sociology of science. *History of Science* 25 (1): 1–70.

Aiken, W. H. 1986. On evaluating agricultural research. In K. A. Dahlberg (ed.), *New Directions for Agriculture and Agricultural Research*. Totawa, NJ: Rowman and Allanheld, pp. 31–41.

Alper, J. 1983. Pollen suppressors open new options for hybrids. *Biol Technology* 1: 14.

Alsberg, C. L., and E. P. Griffing. 1928. The objectives of wheat breeding. *Wheat Studies of the Food Research Institute* 4 (7): 269–88.

American Seed Trade Association (ASTA). 1984. *Position Paper of the American Seed Trade Association on FAO Undertaking on Plant Genetic Resources.* Report of the Public Research Advisory Committee: Subcommittee on Research Priorities. Washington, DC: ASTA.

American Vegetable Grower. 1984. Advances in plant genetics. *American Vegetable Grower* 32 (2): 26.

Anderson, L. A., J. D. Phillipson, and M. F. Roberts. 1985. Biosynthesis of secondary products in cell cultures of higher plants. In A. Fiechter (ed.), *Advances in Biochemical Engineering/Biotechnology*, Plant Cell Culture, no. 31. Berlin: Springer Verlag, pp. 1–36.

Anderson, N. G. 1987. Electrophoresis and large-scale databases. *Science* 235: G65.

Associated Press. 1989. Foreign gene injected into human for first time. *Lexington Herald-Leader*, 23 May, A3.

Atanassova, B., and H. Georgiev. 1986. Investigation of tomato male sterile lines in relation to hybrid seed production. *Acta Horticulturae* 190: 553–7.

Auerbach, B. C. 1983. Biotechnology patent law developments in Great Britain

and the United States: Analysis of a hypothetical patent claim for a synthesized virus. *Boston College International and Comparative Law Review* 6
(Spring): 563–90.

Austin, R. B., R. B. Flavell, I. E. Henson, and H. J. B. Lowe. 1986. *Molecular
Biology and Crop Improvement*. Cambridge: Cambridge University Press.

Avery, O. T., C. M. MacLeod, and M. McCarty. 1944. Studies on the chemical
nature of the substance inducing transformation of pneumococcal types.
Induction of transformation by a desoxyribonucleic acid fraction isolated
from pneumococcus type III. *Journal of Experimental Medicine* 79: 137–
58.

Bailey, L. H. 1886. *Notes on Tomatoes*. Bulletin no. 19. East Lansing: Michigan State Agricultural College.

—— 1887. *Notes on Tomatoes*. Bulletin no. 21. East Lansing: Michigan State
Agricultural College.

Balandrin, M. F., J. A. Klocke, E. S. Wurtele, and W. H. Bollinger. 1985.
Natural plant chemicals: Sources of industrial and medicinal materials.
Science 228: 1154–60.

Baltimore, D. 1982. Priorities in biotechnology. In National Research Council
(ed.), *Priorities in Biotechnology Research for International Development*.
Washington, DC: National Academy Press, pp. 30–7.

Barton, K. A., and W. J. Brill. 1983. Prospects in plant genetic engineering.
Science 219: 671–6.

Bateson, W. 1909. *Mendel's Principles of Heredity*. Cambridge: Cambridge
University Press.

—— 1916. Note on experiments with flax at the John Innes Horticultural
Institution. *Journal of Genetics* 6: 199–201.

Baumgardt, B. R. 1988. Biotechnology and the animal sciences. In W. Lacy and
L. Busch (eds), *Biotechnology and Agricultural Cooperatives: Opportunities
and Challenges*. Lexington, KY: Kentucky Agricultural Experiment Station,
pp. 24–39.

Bawden, F. C., and N. W. Pirie. 1937. The isolation and some properties of
liquid crystalline substances for solanaceous plants infected with three strains
of tobacco mosaic virus. *Proceedings of the Royal Society of London, Series
B* 123: 274–320.

Beattie, J. H. 1921. *Tomatoes for Canning and Manufacturing*. Farmers' Bulletin, no. 1233. Washington, DC: USDA.

Becker, W. M. 1986. *The World of the Cell*. Menlo Park, CA: Benjamin/
Cummings.

Beier, F. K., R. S. Crespi, and J. Straus, 1985. *Biotechnology and Patent
Protection: An International Review*. Paris: Organization for Economic
Cooperation and Development.

Berlan, J.-P. 1987. Recherche sur l'économie politique d'un changement technique: Les myths du maïs hybride. Doctoral dissertation. Les Milles: CEDERS,
University of Aix-Marseille II.

Berlan, J.-P., and R. Lewontin. 1986. Breeders' rights and patenting life forms.
Nature 322: 785–8.

Berlin, J. 1986. Secondary products from plant cell cultures. In H.-J. Rehm and
G. Reed (eds), *Biotechnology: A Comprehensive Treatise in 8 Volumes*.
Weinheim: VCH, vol. 4, pp. 629–58.

Bernal, J. D. 1967. Symmetry of the genetics of form. *Journal of Molecular Biology* 24 (3): 379–90.

Biggle, L. W. 1980. Origin and botany of wheat. In Ernst Hafliger (ed.), *Wheat*. Basel, Switzerland: Ciba-Geigy.

Bijman, J., K. van den Doel, and G. Junne. 1986. *The Impact of Biotechnology on Living and Working Conditions in Western Europe and the Third World*. Dublin: European Foundation for the Improvement of Living and Working Conditions.

Binswanger, H. P. 1978. The microeconomics of induced technical change. In H. P. Binswanger and V. W. Ruttan (eds), *Induced Innovation*. Baltimore: Johns Hopkins University Press, pp. 91–127.

Binswanger, H. P., and V. W. Ruttan. 1978. Introduction. In H. P. Binswanger and V. W. Ruttan (eds), *Induced Innovation*. Baltimore: Johns Hopkins University Press, pp. 1–9.

Binswanger, H. P., and J. G. Ryan. 1979. Village level studies as a locus for research and technology adaptation. Paper presented at International Symposium on Development and Transfer of Technology for Rainfed Agriculture and the SAT Farmer, Hyderabad. Hyderabad: ICRISAT.

Bio/Technology. 1988a. 1989 Bio/Technology buyers guide. *Bio/Technology* 6: 1429–1543.

—— 1988b. In the news. *Bio/Technology* 6: 1366.

—— 1989a. In the news. *Bio/Technology* 7: 202.

—— 1989b. Major mergers, companies combine. *Bio/Technology* 7: 401.

—— 1990. In the news. *Bio/Technology* 8: 90.

Blair, J. M. 1972. *Economic Concentration*. New York: Harcourt Brace Jovanovich.

Blassen, W. 1989. Public concerns regarding applications of biotechnologies. *Biodeterioration Research* 2: 7–17.

Blaxter, K. L. 1975. Concluding remarks. In Arnold Spicer (ed.), *Bread: Social, Nutritional, and Agricultural Aspects of Wheaten Bread*. London: Applied Science Publishers, pp. 339–50.

Bleche, B. 1987. The processing tomato situation in the Mediterranean countries. *Acta Horticulturae* 200: 35–44.

Blumenthal, D., M. Gluck, and S. Epstein, 1987. *University–Industry Relationships in Biotechnology: Implications for Federal Policy*. Bethesda, MD: US Department of Health and Human Services.

Blumenthal, D. M., M. Gluck, K. S. Louis, M. A. Stoto, and D. Wise. 1986a. University–industry research relations in biotechnology: Implications for the university. *Science* 232: 1361–6.

Blumenthal, D. M., M. Gluck, K. S. Louis, and D. Wise. 1986b. Industrial support of university research in biotechnology. *Science* 231: 242–6.

Board on Science and Technology for International Development. 1986. Cotton fibers grown directly from cells. *BOSTID Developments* 6: 12.

Bohme, G. W., R. Van den Daele, W. Hohlfeld, W. Krohn, and W. Schafer. 1983. Introduction. In W. Schafer (ed.), *Finalization in Science: The Social Orientation of Scientific Progress*. Dordrecht: D. Reidel, pp. 3–11.

Bohr, N. 1933. Light and life. *Nature* 131: 421–3, 457–9.

Booth, W. 1989. NIH scientists agonize over technology transfer. *Science* 243: 20–1.

Borlaug, N. 1983. Contributions of conventional plant breeding to food production. *Science* 219: 689–93.

Boswell, V. R. 1937. Improvement and genetics of tomatoes, peppers, and eggplant. *The Yearbook of Agriculture*. Washington, DC: USDA, pp. 176–206.

Boyer, C. D. 1984. Genetic engineering: Tomorrow's technology. *American Vegetable Grower* 32 (4): 51.

Bragg, W. 1913. The structure of some crystals as indicated by their diffraction of X-rays. *Proceedings of the Royal Society of London, Series A* 89: 248–77.

Brandt, J. A., B. C. French, and E. V. Jesse. 1978. *Economic Performance of the Processing Tomato Industry*. Bulletin 1888, Giannini Foundation of Agricultural Economics Information Series no. 78–1. Davis, CA: University of California.

Brannigan, A. 1981. *The Social Basis of Scientific Discoveries*. Cambridge: Cambridge University Press.

Breathnach, R., J. L. Mandell, and P. Chambon. 1977. Ovalbumin gene is split. *Nature* 270: 314–19.

Breuling, M., A. W. Alfermann, and E. Reinhard. 1985. Cultivation of cell cultures of *Berberis wilsonae* in 20–1 airlift bioreactors. *Plant Cell Reports* 4: 220–3.

Brill, W. J. 1985. Safety concerns and genetic engineering in agriculture. *Science* 227: 381–4.

Brockway, L. H. 1979. *Science and Colonial Expansion: The Role of the British Royal Botanic Gardens*. New York: Academic Press.

Brooks, H. J., and G. Vest. 1985. Public programs on genetics and breeding of horticultural crops in the United States. *Hort Science* 20 (October): 826–30.

Buller, A. H. R. 1919. *Essays on Wheat*. New York: Macmillan.

Burkhardt, J. 1986. Agribusiness ethics: Specifying the terms of the contract. *Journal of Business Ethics* 5 (3): 333–45.

Burmeister, L. L. 1987. The South Korean green revolution: Induced or directed innovation? *Economic Development and Cultural Change* 35: 767–90.

Busch, L. 1978. On understanding understanding: Two views of communication. *Rural Sociology* 43 (3): 450–73.

—— 1980. Structure and negotiation in the agricultural sciences. *Rural Sociology* 45 (1): 26–48.

—— 1984. Science, technology, agriculture, and everyday life. *Research in Rural Sociology and Development* 1: 289–314.

Busch, L., and W. B. Lacy. 1983. *Science, Agriculture, and the Politics of Research*. Boulder, CO: Westview Press.

—— (eds). 1984. *Food Security in the United States*. Boulder, CO: Westview Press.

Busch, L., W. B. Lacy, and J. Burkhardt. 1988. Culture and care: Ethical and policy dimensions of germplasm conservation. Paper presented at Beltsville Symposium XIII, Biotic Diversity and Germplasm Conservation, Beltsville, MD.

Busch, L., J. L. Silver, W. B. Lacy, C. S. Perry, M. Lancelle, and S. Deo. 1984. *The Relationship of Public Agricultural R&D to Selected Changes in the Farm Sector*. A report to the National Science Foundation. Lexington, KY: University of Kentucky, Department of Sociology.

Buttel, F. H. 1981. *American Agriculture and Rural America: Challenges for Progressive Politics*. Bulletin no. 120. Ithaca, NY: Cornell University, Department of Rural Sociology.

Buttel, F. H., and R. Barker. 1985. Emerging agricultural biotechnologies, public policy, and implications for Third World agriculture: The case of biotechnology. *American Journal of Agricultural Economics* 67: 1170–5.

Buttel, F. H., and J. Belsky. 1987. Biotechnology, plant breeding and intellectual property: Social and ethical dimensions. *Science, Technology, and Human Values* 12 (1): 31–49.

Buttel, F. H., M. Kenney, J. Kloppenburg Jr., D. Smith, and J. T. Cowan. 1986. Industry/land grant university relationships in transition. In L. Busch and W. B. Lacy (eds), *The Agricultural Scientific Enterprise: A System in Transition*. Boulder, CO: Westview Press, pp. 296–312.

Byé, P., and A. Mounier. 1984. Les futurs alimentaires et énergetiques des biotechnologies. *Economies et Sociétés*, Hors Serie no. 27.

Cairns, J., G. Stent, and J. Watson (eds). 1966. *Phage and the Origins of Molecular Biology*. Cold Spring Harbor, NY: Cold Spring Harbor Laboratory for Quantitative Biology.

Calado, A. M., C. A. M. Portas, and M. J. R. Figueiredo. 1980. Growth and development of tomato "Cal j". *Acta Horticulturae* 100: 159–71.

California Agricultural Lands Project. 1982. *Quick Book: Genetic Engineering of Plants*. San Francisco: California Agricultural Lands Project.

Canadian Department of Agriculture. 1967. *Canada Agriculture: The First Hundred Years*. Ottawa: The Queen's Printer.

Carleton, M. A. 1900. *The Basis for the Improvement of American Wheats*. Bulletin no. 24. Washington, DC: USDA, Division of Vegetable Physiology and Pathology.

Carlson, E. A. 1966. *The Gene: A Critical History*. Philadelphia: W. B. Saunders.

—— 1971. An unacknowledged founding of molecular biology: H. J. Muller's contributions to gene theory. *Journal of the History of Biology* 4: 149–70.

—— 1974. Hermann Joseph Muller. In C. C. Gillespie (ed.), *Dictionary of Scientific Biography*, vol. 9. New York: Scribner p. 565.

Carpenter, K. J. 1975. The nutritive value of wheat proteins. In A. Spicer (ed.), *Bread: Social, Nutritional, and Agricultural Aspects of Wheaten Bread*. London: Applied Science Publishers, pp. 93–114.

Casili, V. W. D., and E. C. Tigchelaar. 1975. Computer simulation studies comparing pedigree, bulk and single seed descent selection in self-pollinated populations. *Journal of the American Society of Horticultural Science* 100: 364–7.

Caspersson, T., and J. Schultz. 1938. Nucleic acid metabolism of the chromosomes in relation to gene production. *Nature* 142: 294–5.

Cass, M. 1990. Japanese seem headed toward ownership of US Companies. *Genetic Engineering News* 10 (1): 1.

Centre Internacional de Mejoramiento de Maiz Y Trigo (CIMMYT). 1985. *CIMMYT World Wheat Facts and Trends*. Report Three: A Discussion of Selected Wheat Marketing and Pricing Issues in Developing Countries. Mexico City: CIMMYT.

Chaleff, R. S. 1983. Isolation of agronomically useful mutants from plant cell cultures. *Science* 219: 676–82.

Chamala, S. 1985. Transfer of rubber technology among smallholders in Malaysia and Indonesia: A sociological analysis. *Smallholder Rubber Production and Policies*. Canberra: Australian Centre for International Agricultural Research, pp. 30–4.

Chamberlain, N. 1975. Advances in breadmaking technology. In A. Spicer (ed.), *Bread: Social, Nutritional, and Agricultural Aspects of Wheaten Bread*. London: Applied Science Publishers, pp. 259–72.

Chang, J. C., and Y. W. Kan. 1981. Antenatal diagnosis of sickle-cell anemia by direct analysis of the sickle cell mutation. *Lancet* 2: 1127–9.

Chargaff, E. 1951. Structure and function of nucleic acids as cell constituents. *Federation Proceedings* 10: 654–9.

Charvet, J. P. 1988. *La Guerre du Blé*. Paris: Economica.

Chatelin, Y. 1986. La science et le développement: L'Histoire peut-elle recommencer? *Revue Tiers Monde* 27: 5–24.

Chatelin, Y., and R. Arvanitis. 1984. Pratiques et Politiques Scientifiques. Actes du Forum des 6 et 7 février. Paris: ORSTOM.

Chicago Board of Trade. 1982. *Grains: Production, Processing, Marketing*. Chicago: Chicago Board of Trade.

Chilton, M.-D., M. H. Drummond, D. J. Merlo, D. Sciaky, A. C Montoya, M. P. Gordon, and E. W. Nester. 1977. Stable incorporation of plasmid DNA into higher plant cells: The molecular basis of crown gall tumorigenesis. *Cell* 11: 263–71.

Cisar, G. 1983. Chemically assisted hybrid production. Paper presented at the Southern Small Grain Workers Conference, Fayetteville, AK, 5–7 April.

Clairmonte, F., and J. Cavanaugh. 1986. Destruction of the sugar industry. *Economic and Political Weekly* 21 (Jan. 4): 18–19.

Cocking, E. C. 1983. Hybrid and cybrid production via protoplast fusion. In L. D. Woens (ed.), *Genetic Engineering: Applications to Agriculture*. Totowa, NJ: Rowman and Allanheld, pp. 257–68.

—— 1986. A case study of the application of protoplast fusion to tomato improvement. In P. A. Day (ed.), *Biotechnology and Crop Improvement and Protection*. Thornton Heath: British Crop Protection Council, pp. 43–8.

Cohen, S., A. Chang, H. Boyer, and R. Helling. 1973. Construction of biologically functional bacterial plasmids *in vitro*. *Proceedings of the National Academy of Sciences* 70: 3240–4.

Coleman, W. 1970. Bateson and chromosomes: Conservative thought in science, *Centaurus* 15: 228–314.

Colin, H. A., and M. Watts. 1983. Flavor production in culture. In D. A. Evans, W. R. Sharp, P. V. Ammirato, and Y. Yamada (eds), *Handbook of Plant Cell Culture*, vol. 1. New York: Macmillan, pp. 729–47.

Columella, L. 1911. *On Agriculture*. Trans. by E. S. Forster and E. H. Heffner. 3 vols, Cambridge: Harvard University Press.

Comis, D., and M. Wood. 1986. Assembly line plants take root. *Agricultural Research* 34 (4): 6–11.

Corbett, L. C. 1905. *Tomatoes*. Farmers' Bulletin, no. 220. Washington, DC: USDA.

Council for Agricultural Science and Technology (CAST). 1985. *Plant Germplasm Preservation and Utilization in US Agriculture*. CAST Report, no. 106. Ames, Iowa: CAST.

Crick, F. H. C. 1958. On protein synthesis. *Symposia of the Society for Experimental Biology* 12: 548–55.

Curtin, M. E. 1983. Harvesting profitable products from plant tissue culture. *Bio/Technology* 1: 649–52, 654, 656–7.

Dahlberg, K. A. 1986. Introduction: Changing contexts and goals and the need for new evaluative approaches. In K. A. Dahlberg (ed.), *New Directions for Agriculture and Agricultural Research*.Totawa, NJ: Rowman and Allanheld, pp. 1–27.

Daly, P. 1985. *The Biotechnology Business: A Strategic Analysis*. London: Frances Pinter.

Davenport, C. 1981. Sowing the seeds – research, development flourish at DeKalb, Pioneer Hy-bred. *Barron's* 61 (9): 9–10, 33.

Davis, J. S. 1940. Bulk handling in Australia. *Wheat Studies of the Food Research Institute* 16 (7): 301–64.

DeGreve, H., J. Leemans, J.-P. Hernalsteens, L. Thia-Toong, M. DeBeuckeleer, L. Willmitzer, L. Otten, M. VanMontagu, and J. Schell. 1982. Regeneration of normal and fertile plants that express octopine synthase, from tobacco crown galls after deletion of tumor-controlling functions. *Nature* 300: 752–5.

DeJanvry, A., and P. LeVeen. 1983. Aspects of the political economy of technical change in developed economies. In M. Pineiro and E. Trigo (eds), *Technical Change and Social Conflict in Agriculture: Latin American Perspectives*. Boulder, CO: Westview Press.

Dembo, D., C. Dias, and W. Morehouse. 1985. The biorevolution and the Third World. *Third World Affairs* 1: 311–25.

de Oliveira, W. 1980. The production of tomatoes in Portugal. *Acta Horticulturae* 100: 93–8.

Descartes, R. [1637] 1956. *Discourse on Method*. Indianapolis: Bobbs-Merrill.

Dibner, M. D. 1986. Biotechnology in Europe. *Science* 232: 1367–72.

Dibner, M. D., and C. C. Lavich. 1987. An analysis of partnerships. *Bio/Technology* 5: 1029–33.

Dickson, D. 1989. German biotech firms flee regulatory climate. *Science* 244: 1251–2.

Dixon, B. 1984. Chemicals from algae and plant tissue. *Bio/Technology* 2: 665.

—— 1989. Biotech's major impact yet a decade away. *Bio/Technology* 7: 420.

Donaldson, T. 1983. *Corporations and Morality*. Englewood Cliffs, NJ: Prentice-Hall.

Dontlinger, P. T. 1908. *The Book of Wheat: An Economic History and Practical Manual of the Wheat Industry*. New York: Orange Judd.

Dotlacil, L., and M. Apltauerova. 1978. Pollen sterility induced by Ethrel and its utilization in hybridization of wheat. *Euphytica* 27: 353–60.

Doyle, J. 1985. *Altered Harvest: Agriculture, Genetics, and the Future of the World's Food Supply*. New York: Viking Press.

East, E. M. 1912. A Mendelian interpretation of variation that is apparently continuous. *American Naturalist* 46: 633–95.

East, E. M., and D. F. Jones. 1920. Genetic studies on the protein content of maize. *Genetics* 5: 543–610.

Eggitt, P., W. Russell, and A. W. Hartley. 1975. Extraction rates – milling processing implications. In A. Spicer (ed.), *Bread: Social, Nutritional, and Agricultural Aspects of Wheaten Bread*. London: Applied Science Publishers, pp. 219–33.

Elster, J. 1983. *Explaining Technical Change*. Cambridge: Cambridge University Press.

Elton, G. A. H. 1975. The fortification of flour. In A. Spicer, (ed.), *Bread: Social, Nutritional, and Agricultural Aspects of Wheaten Bread*. London: Applied Science Publishers, pp. 249–57.

European Economic Community, Directorate General for Research. 1986. *Synopsis of the Hearing on Biotechnology*, 14 January. Brussels: EEC.

Evans, D. A., and W. R. Sharp. 1983. Single gene mutations in tomato plants regenerated from tissue culture. *Science* 221: 949–51.

—— 1986. Potential applications of plant cell culture. In S. K. Harlander and T. P. Labuza (eds), *Biotechnology in Food Processing*. Park Ridge, NJ: Noyes Publications, pp. 133–43.

Evans, D. A., W. R. Sharp, and H. P. Medina-Filho. 1984. Somaclonal and gametoclonal variation. *American Journal of Botany* 71 (6): 759–74.

Evenson, D. D., and R. E. Evenson. 1983. Legal systems and private sector initiatives for the invention of agricultural technology in Latin America. In M. Pineiro and E. Trigo (eds), *Technical Change and Social Conflict in Agriculture: Latin American Perspectives*. Boulder, CO: Westview Press, pp. 189–233.

Federal Register. 1986. Advance notice of proposed USDA regulations for biotechnology research. *Federal Register* S1: 23367.

Fejer, S. O. 1966. The problem of plant breeders' rights. *Agricultural Science Review*, Third Quarter, 1–7.

Ferguson, E. S. 1980. *Oliver Evans: Inventive Genius of the American Industrial Revolution*. Greenville, DE: The Hagley Museum.

Feyerabend, P. 1975. *Against Method*. Atlantic Highlands, NJ: Humanities Press.

—— 1978. *Science in a Free Society*. London: New Left Books.

Fillatti, J. J., J. Kiser, B. Rose, and L. Comai. 1987. Efficient transformation of tomato and the introduction and expression of a gene for herbicide tolerance. In D. J. Nevins and R. A. Jones (eds), *Tomato Biotechnology*. New York: Alan R. Liss, pp. 199–210.

Fisher, A. 1986. Science newsfront. *Popular Science* 228 (5): 10, 14.

Fishhoff, D. A., K. S. Bowdesh, F. J. Perlak, P. G. Marione, S. M. McCormick, J. G. Niedermeyer, D. A. Dean, K. Kusano-Kretzmer, E. J. Mayer, D. E. Rochester, S. G. Rogers, and R. T. Fraley. 1987. Insect tolerant transgenic tomato plants. *Bio/Technology* 5: 807–13.

Flavell, R. B. 1985. Molecular and biochemical techniques in plant breeding. In *Biotechnology and International Agricultural Research: Proceedings of the Inter-Center Seminar on International Agricultural Research Centers and Biotechnology*. Manila: International Rice and Research Institute, pp. 247–56.

Flora, J. 1986. History of wheat research at the Kansas Agricultural Experiment Station. In L. Busch and W. B. Lacy (eds), *The Agricultural Scientific Enterprise: A System in Transition*. Boulder, CO: Westview Press, pp. 186–205.

Fontanel, A., and M. Tabata. 1987. Production of secondary metabolites by plant tissue and cell cultures: Present aspects and prospects. *Nestle Research News*, pp. 93–103.

Food and Agriculture Organization. 1985. *FAO Trade Yearbook, 1984*. Rome: FAO.

Foucault, M. 1970. *The Order of Things*. New York: Random House.

Fowler, C., E. Lachkovics, P. Mooney, and H. Shand. 1988. The laws of life: Another development and the new biotechnologies. *Development Dialogue* 1–2: 1–350.

Fowler, M. W. 1984. Large-scale cultures of cells in suspension. In I. K. Vasil (ed.), *Cell Culture and Somatic Cell Genetics of Plants*. New York: Academic Press, pp. 167–74.

—— 1985. Plant cell culture – Future perspectives. In K. H. Neuman, W. Barz, and E. Reinheld (eds), *Primary and Secondary Metabolites of Plant Cell Cultures*. Berlin: Springer Verlag, pp. 362–70.

—— 1988. Problems in commercial exploitation of plant cell cultures. In G. Bock and J. Marsh (eds), *Applications of Plant Cell and Tissue Culture*. Ciba Foundation Symposium no. 137. Chichester: John Wiley and Sons, pp. 239–53.

Fox, J. L. 1989. Debate continues on patenting organisms. *Bio/Technology* 7: 414.

—— 1990. Reassessing the scope of federal biotechnology issues. *ASM News*. 56 (3): 142–3.

Fraley, R. 1989. Biotechnology on the farm: Impact of new technologies on crop production. Paper presented at the annual meeting of the American Association for the Advancement of Science, San Francisco, CA.

Franklin, R. E., and R. Gosling. 1953. Molecular configuration in sodium thymonucleate. *Nature* 171: 740–1.

French, P. 1979. Corporate moral agency. In T. Brauchamp and N. Bowie (eds), *Ethical Theory and Business*. Englewood Cliffs, NJ: Prentice-Hall, pp. 175–86.

Friedland, W. H. 1984. The labor force in US agriculture. In L. Busch and W. B. Lacy (eds) *Food Security in the United States*. Boulder, CO: Westview Press, pp. 143–82.

—— 1985. Institutional hegemony and the malformation of agricultural science. Paper presented at the annual meeting of the American Association for the Advancement of Science, Los Angeles, CA.

Friedland, W. H., and A. Barton. 1975. *Destalking the Wily Tomato*. Research Monograph no. 15. Davis, CA: University of California, Department of Applied Behavioral Sciences.

Friedland, W. H., A. E. Barton, and R. J. Thomas. 1981. *Manufacturing Green Gold: Capital, Labor, and Technology in the Lettuce Industry*. Cambridge: Cambridge University Press.

Friedland, W. H., and T. Kappel. 1979. *Production or Perish: Changing the*

Inequities of Agricultural Research Priorities. Santa Cruz, CA: Project on Social Impact Assessment and Values, University of California.

Friedman, M. 1962. *Capitalism and Freedom.* Chicago: University of Chicago Press.

Friedmann, H. 1978. World market, state, and family farm: Social bases of household production in the era of wage labor. *Comparative Studies in Society and History* 20 (4): 545–86.

—— 1982. The political economy of food: The rise and fall of the postwar international food order. *American Journal of Sociology* 88 (Supp.): S248–86.

Fuchsberg, G. 1989. Cornell University to step up efforts to market its research. *Chronicle of Higher Education,* 35 (39): A29–30.

Fujita, Y. 1988. Industrial production of shikonin and berberine. In G. Bock and J. Marsh (eds), *Applications of Plant Cell and Tissue Culture.* Ciba Foundation Symposium No. 137. Chichester: John Wiley and Sons, pp. 228–38.

Gardner, B. A. 1981. *The Governing of Agriculture.* Lawrence, KS: University of Kansas Press.

Gasser, C. S., and R. T. Fraley. 1989. Genetically engineering plants for crop improvement. *Science* 244: 1293–9.

Gates, P. W. 1960. *The Farmer's Age: Agriculture, 1815–1860.* New York: Harper and Row.

Gautheret, R. J. 1983. Plant tissue culture: A history. *The Botanical Magazine* (Tokyo) 96: 393–410.

Gebhart, F. 1990. Industry considers the Roche–Genentech deal a major vote of confidence for biotechnology. *Genetic Engineering News* 10 (3): 1, 22.

Genetic Engineering News. 1984. 1984 Guide to Biotechnology Companies. *Genetic Engineering News* 4 (10): 15–37.

—— 1985. Fourth annual GEN guide to biotechnology companies. *Genetic Engineering News* 5 (10): 24–47.

—— 1987. Sixth annual GEN guide to biotechnology companies. *Genetic Engineering News* 7 (10): 21–54.

—— 1990a. New plant hybrid tool developed. *Genetic Engineering News* 10 (1): 1, 38.

—— 1990b. Plant Science Research, Inc. achieves successful transformation of corn. *Genetic Engineering News* 10 (3): 3.

George, W. L., Jr., and S. A. Berry. 1983. Genetics and breeding of processing tomatoes. In W. A. Gould (ed.), *Tomato Production, Processing and Quality Evaluation.* 2nd ed. Westport, CT: AVI, pp. 48–65.

Gerlach, W. L., E. S. Dennis, and W. J. Peacock. 1985. Approaches to the transformation of crop plants. *Biotechnology in International Agricultural Research: Proceedings of the Inter-Center Seminar on International Agricultural Research Centers and Biotechnology.* Manila: International Rice Research Institute, pp. 257–67.

Gibbs, J. N., and J. S. Kahan. 1986. Federal regulation of food and food additive biotechnology. *Administrative Law Review* 38 (Winter): 1–32.

Giedion, S. 1975. *Mechanization Takes Command.* New York: W. W. Norton.

Gill, P. R., and G. J. Warren. 1988. An iron-antagonized fungistatic agent that is not required for iron assimilation from a fluorescent rhizophere pseudomonad. *Journal of Bacteriology* 170: 163–70.

Glick, M. 1988. New provost inaugurates ag bioethics newsletter. *Ag Bioethics Forum* 1 (1): 1.

Goldberg, R. A. 1968. *Agribusiness Coordination: A Systems Approach to the Wheat, Soybeans, and Florida Orange Economies.* Boston: Harvard University Graduate School of Business Administration.

—— 1983. Economies of the Flour Milling Industry. *Research in Domestic and International Agribusiness Management* 4: 145–95.

Goldschmidt, R. 1916. Genetic factors and enzyme reaction. *Science* 43: 98–100.

Goldstein, D. J. 1986. Molecular biology and the preservation of germplasm. Paper presented at meeting of the American Association for the Advancement of Science, Philadephia, May 1986.

Goldstein, W. E. 1985. Large-scale processing of plant cell culture. *Annals of the New York Academy of Sciences* 413: 394–408.

Goodman, D., B. Sorj, and J. Wilkinson. 1987. *From Farming to Biotechnology: A Theory of Agro-Industrial Development.* Oxford: Basil Blackwell.

Goonatilake, S. 1982. *Crippled Minds: An Exploration into Colonial Culture.* New Delhi: Vikas.

Gould, W. A. 1983. *Tomato Production, Processing and Quality Evaluation.* 2nd ed. Westport, CT: AVI.

Grall, J. 1986. Les moyens du miracle. *Croissance des Jeunes Nations* 287 (10): 47–50.

Greenberg, D. S. 1966. Bootlegging: It holds a firm place in conduct of research. *Science* 153: 848–9.

Gussow, J. D. 1984. Food Security in the United States: a nutritionist's viewpoint. In L. Busch and W. B. Lacy (eds), *Food Security in the United States.* Boulder, CO: Westview Press, pp. 207–30.

Habermas, J. 1971. *Knowledge and Human Interests.* Boston: Beacon Press.

Hadwiger, D. F. 1970. *Federal Wheat Commodity Programs.* Ames, IA: Iowa State University Press.

Hagen, K. J. 1973. Showing the flag in the Indian ocean. In C. Barrow, Jr. (ed.), *America Spreads Her Sails.* Annapolis: Naval Institute Press, pp. 153–75.

Hall, B. D., and S. Spiegelman. 1961. Sequence complementarity of T2 DNA and T2-specific RNA. *Proceedings of the National Academy of Sciences* 47: 137–46.

Hanauer, S. W. 1900. Tomatoes and cantaloupes in Europe. *Consular Reports* 64 (11): 394–5.

Hansen, M., L. Busch, J. Burkhardt, W. B. Lacy, and L. R. Lacy. 1986. Plant breeding and biotechnology. *BioScience* 36 (1): 29–39.

Harlander, S. K. 1989. Innovations and challenges in food biotechnology. Paper presented at the meeting of the Institute of Food Technologists, Chicago, IL.

Hawkes, J. G. 1983. *The Diversity of Crop Plants.* Cambridge, MA: Harvard University Press.

Hawley, B. 1984. *Statement of the American Farm Bureau Federation to the Subcommittee on Agricultural Research and General Legislation of the Senate Agriculture, Nutrition and Forestry Committee Regarding S. 1128 and H. R.*

2714 The Agricultural Productivity Act of 1983. Washington, DC: American Farm Bureau Federation.

Hebblethwaite, J. F. 1988. The future of plant biotechnology in agriculture: A corporate strategy. In W. B. Lacy and L. Busch (eds), *Biotechnology and Agricultural Cooperatives: Opportunities and Challenges,* Lexington, KY: Kentucky Agricultural Experiment Station, pp. 66–70.

Heid, W. G. 1979. *US Wheat Industry.* Agricultural Economic Report, no. 432. Washington, DC: USDA, Economics, Statistics, and Cooperatives Service.

Heinstein, P. F. 1985. Future approaches to the formation of secondary natural products in plant cell suspension cultures. *Journal of Natural Products* 48: 1–9.

Hicks, J. R. 1932. *The Theory of Wages.* London: Macmillan.

Hightower, J. 1973. *Hard Tomatoes, Hard Times.* Cambridge, MA: Schenckman.

Hildebrand, P. E. 1980–1. Motivating small farmers, scientists and technicians to accept change. *Agricultural Administration* 8: 375–83.

Hollis, M., and E. J. Nell. 1975. *Rational Economic Man.* Cambridge: Cambridge University Press.

Hughes, W. G., M. D. Bennett, J. J. Bodden, and S. Galanopoulou. 1974. Effects of time of application of Ethrel on male sterility and ear emergence in wheat. *Triticum aestivum. Annals of Applied Biology* 76: 243–52.

Hughes, W. G., and J. J. Bodden. 1977. Single-gene restoration of cytoplasmic male sterility in wheat and its implications in the breeding of restorer lines. *Theoretical and Applied Genetics* 50: 129–35.

Idhe, D. 1979. *Technics and Praxis.* Dordrecht: D. Reidel.

Ilker, R. 1987. In-vitro pigment production: An alternative to color synthesis. *Food Technology* 41 (4): 70–72.

Industries Assistance Commission (IAC). 1978. *Wheat Stabilization.* Report No. 175. Canberra: Australian Government Publishing Service.

—— 1988. *The Wheat Industry.* Report No. 411. Canberra: Australian Government Publishing Service.

Irvine, G. N. 1975. Milling and baking quality. In A. Spicer (ed.), *Bread: Social, Nutritional, and Agricultural Aspects of Wheaten Bread.* London: Applied Science Publishers, pp. 114–26.

James, A. T. 1984. Plant tissue culture: Achievements and prospects. *Proceedings of the Royal Society of London Series B* 222: 134–45.

Jan, C. C., and P. L. Rowell. 1981. Response of wheat tillers at different growing stages to gametocide treatment. *Euphytica* 30: 501–4.

Jesse, E. V., W. G. Schultz, and J. L. Bomben. 1975. *Decentralized Tomato Processing: Plant Design, Costs, and Economic Feasibility.* Agricultural Economic Report, no. 313. Washington, DC: USDA, Economic Research Service.

Joergensen, J. 1951. The development of logical empiricism. In A. Neurath, R. Carnap, and C. Morris (eds). *International Encyclopedia of Unified Science,* vol 2. Chicago: University of Chicago Press, pp. 845–946.

Johnston, R. A. 1985. Hybrid wheat economics: Looking better all the time. *Crops and Soils* 37 (9): 15–18.

Jonas, H. 1984. *The Imperative of Responsibility.* Chicago: University of Chicago Press.

Jones, R. A., and S. J. Scott. 1983. Improvement of tomato flavor by genetically increasing sugar and acid contents. *Euphytica* 32: 845–55.

Jorgensen, R. A. 1987. A hybrid seed production method based on synthesis of novel linkages between marker and male-sterile genes. *Crop Science* 27: 806–10.

Jost, M., M. Glatki-Jost, and V. Hrust. 1975. Influence of *T. timopheevi* cytoplasm on characteristics of male sterile common wheat 3. The plant morphology and kernel characteristics. *Cereal Research Communications* 3: 15–26.

Kalton, R. 1984. Who is going to breed the minor crops and train the next generation of plant breeders? *Seed World* 122 (10): 16–17.

Kannenberg, L. W. 1984. Utilization of genetic diversity in crop breeding. In C. W. Yeatman, D. Kafton and G. Wilkes (eds), *Plant Genetic Resources: A Conservation Imperative*. Boulder, CO: Westview Press, pp. 93–109.

Kao, K. N. 1978. Plant protoplast fusion and somatic hybridization. *In Proceedings of the Symposium on Plant Tissue Culture*, 1978 May 25–30, Bejing, China: Science Press, pp. 331–9.

Kendrew, J. C., G. Bodo, H. M. Dintzis, R. J. Parrish, H. Wyckoff, and D. C. Philipps. 1958. A three-dimensional model of the myoglobin molecule obtained by X-ray analysis. *Nature* 181: 662–6.

Kenney, M. 1986. *Biotechnology: The University–Industrial Complex*. New Haven: Yale University Press.

Kenney, M., F. H. Buttel, J. T. Cowan, and J. Kloppenburg, Jr. 1982. *Genetic Engineering and Agriculture: Exploring the Impacts of Biotechnology on Industrial Structure, Industry-University Relationships, and the Social Organization of US Agriculture*. Bulletin no. 125. Ithaca, NY: Cornell University, Department of Rural Sociology.

Kenney, M., F. H. Buttel, and J. Kloppenburg, Jr. 1984. Impact of industrial applications: Socioeconomic impact of product dislocations. *Advance Technology Alert Bulletin* 1: 48–51.

Kindleberger, C. P. 1951. Group behavior and international trade. *Journal of Political Economy* 59 (1): 30–46.

Kingswood, K. 1975. The structure and biochemistry of the wheat grain. In A. Spicer (ed.), *Bread: Social, Nutritional, and Agricultural Aspects of Wheaten Bread*. London: Applied Science Publishers, pp. 47–65.

Klassen, W. 1989. Public concerns regarding applications of biotechnologies. *Biodeterioration Research* 2: 7–17.

Klausner, A. 1986. Gains, hardships to stem from agbiotech. *Bio/Technology* 4: 385.

—— 1987. Doing business with Japan. *Bio/Technology* 5: 1019, 1023–4, 1026.

—— 1989. Biotech changing agribusiness. *Bio/Technology* 7: 219.

Klein, T. M., E. C. Harper, Z. Svab, J. C. Sanford, M. E. Fromm, and P. Maliga. 1988. Stable genetic transformation of intact *Nicotiana* cells by the particle bombardment process. *Proceedings of the National Academy of Sciences* 85 (22): 8502–5.

Kloppenburg, J. 1984. The social impacts of biogenetic technology in agriculture: Past and future. In G. Berardi and C. Geisler (eds), *The Social Consequences and Challenges of the New Agricultural Biotechnologies*. Boulder, CO: Westview Press, pp. 291–321.

—— 1987. *First the Seed: The Political Economy of Plant Biotechnology, 1492–2000.* New York: Cambridge University Press.

Knorr, D., and K. L. Clancy. 1984. Safety aspects of processed foods. In L. Busch and W. B. Lacy (eds), *Food Security in the United States.* Boulder, CO: Westview Press, pp. 231–53.

Knorr-Cetina, K. D. 1981. *The Manufacture of Knowledge.* Oxford: Pergammon Press.

Kohler, R. E. 1978. A Policy for the Advancement of Science: The Rockefeller Foundation, 1924–29. *Minerva* 16: 480–515.

—— 1980. Warren Weaver and the Rockefeller Foundation program in molecular biology: A case study in the management of science. In Nathan Reingold (ed.), *Sciences in the American Context.* Washington, DC: Smithsonian Institution Press, pp. 249–93.

Koornneef, M., C. Hanhart, M. Jongsma, I. Toma, R. Weide, P. Zabel, and J. Hille. 1986. Breeding of a tomato genotype readily accessible to genetic manipulation. *Plant Science* 45: 201–8.

Kornberg, A. 1960. Biological synthesis of deoxyribonucleic acid. *Science* 131: 503–8.

Krimsky, S. 1984. Corporate academic ties in biotechnology: A report on research in progress. *Gene Watch*, September–December, 3–5.

Krohn, W., and W. Schafer. 1983. Agricultural chemistry: The origin and structure of a finalized science. In W. Schafer (ed.), *Finalization in Science: The Social Orientation of Scientific Progress.* Dordrecht: D. Reidel, pp. 17–52.

Kuhn, T. 1970. *The Structure of Scientific Revolutions.* Chicago: University of Chicago Press.

Kunimoto, L. 1986. Commercial opportunities in plant biotechnology for the food industry. *Food Technology* 40 (10): 58–60.

Lacy, W. B., and L. Busch. 1988. Biotechnology: Consequences and strategies for cooperatives. In W. Lacy and L. Busch (eds), *Biotechnology and Agricultural Cooperatives: Opportunities and Challenges.* Lexington, KY, Kentucky Agricultural Experiment Station, pp. 83–105.

—— 1989. Changing division of labor between the university and industry: The case of agricultural biotechnology. In J. Molnar and H. Kinnucan (eds), *Biotechnology and the New Agricultural Revolution.* American Association for the Advancement of Science Symposium Series. Boulder, CO: Westview Press, pp. 21–50.

Lacy, W. B., L. R. Lacy, and L. Busch. 1988. Agricultural biotechnology research: Practices, consequences and policy recommendations. *Agriculture and Human Values* 5 (3): 3–14.

Lakatos, I. 1970. Falsification and the methodology of scientific research programs. In I. Lakatos and R. Musgrave (eds), *Criticism and the Growth of Knowledge.* Cambridge: Cambridge University Press, pp. 91–196.

Latour, B. 1983. Give me a laboratory and I will raise the world. In K. Knorr-Cetina and M. Mulkay (eds). *Science Observed.* London: Sage Publications, pp. 141–70.

—— 1984. *Les Microbes: Guerre et Paix suivi de Irréductions.* Paris: Editions A. M. Métailié.

—— 1986. Visualization and cognition: Thinking with eyes and hands. *Knowledge and Society* 6: 1–40.

—— 1987. *Science in Action: How to Follow Scientists and Engineers Through Society*. Milton Keynes, England: Open University Press.

Latour, B., and S. Woolgar. 1979. *Laboratory Life: The Social Construction of Scientific Facts*. Beverly Hills: Sage Publications.

Laudan, R. (ed.). 1984. *The Nature of Technological Knowledge: Are Models of Scientific Change Relevant?* Dordrecht: D. Reidel.

Lawson, C., W. Kaniewski, L. Haley, R. Rozman, C. Newell, P. Sanders, and N. Tumer. 1990. Engineering resistance to mixed virus infection in a commercial potato cultivar: resistance to potato virus X and potato virus Y in transgenic Russet Burbank. *Bio/Technology* 8 (2): 127–43.

Lehrman, S. 1989. Biotechnology: Man-made miracles. *San Francisco Examiner*, 15 January, 1, 12.

Lenin, V. I. [1907] 1938. *Theory of the Agrarian Question*. In *V. I. Lenin, Selected Works*, vol. 12. New York: International Publishers.

Leonard, E. D. 1987. The prospective impact of biotechnology on the fats and oils industry. *Soap/Cosmestics/Chemical Specialties Journal* 63 (November): 33–4, 81.

Levins, R. 1973. Fundamental and applied research in agriculture. *Science* 181: 523–4.

Levins, R., and R. Lewontin. 1985. *The Dialectical Biologist*. Cambridge, MA: Harvard University Press.

Levitt, T. 1958. The dangers of social responsibility. *Harvard Business Review* 36 (5): 41–50.

Lindow, S. E., N. J. Panopoulous, and B. L. McFarland. 1989. Genetic engineering of bacteria from managed and natural habitats. *Science* 244: 1300–7.

Loffler, C. M., R. N. Busch, and J. V. Wiersma. 1983. Recurrent selection for grain protein percentage in hard red spring wheat. *Crop Science* 23: 1097–1101.

Mahlin, J. C. [1944] 1973. *Winter Wheat in the Golden Belt of Kansas*. New York: Octagon Books.

Makulowick, J. S. 1988. Merger of DNAP and AGS weds experience in plant tissue culture to recombinant DNA skills. *Genetic Engineering News* 8 (2): 1,33.

Maluf, W. R. 1986. Tomato breeding in mild winter tropical areas of Brazil: Goals and trends. *Acta Horticulturae* 191: 331–40.

Mannheim, K. 1936. *Ideology and Utopia*. New York: Harcourt, Brace and Co.

Margolis, J. 1978. Conflict of interest and conflicting interest. In T. Brauchamp and N. Bowie (eds), *Ethical Theory and Business*. Englewood Cliffs, NJ: Prentice-Hall, pp. 361–72.

Marshall, D. R., and A. H. D. Brown. 1981. Wheat genetic resources. In L. T. Evans and W. J. Peacock (eds), *Wheat Science – Today and Tomorrow*. Cambridge: Cambridge University Press, pp. 21–40.

Martin, C. C. 1990. Japanese bioindustry trends turn into firmly established strategies. *Genetic Engineering News* 10 (2): 20.

Martin-Leake, H. 1975. An historical memoir of the indigo industry in Bihar. *Economic Botany* 29: 361–71.

Matthiessen, C., and H. Kohn. 1984. In search of the perfect tomato. *The Nation*, 7–14 July.

Maunder, P. [1970?]. *The Bread Industry in the United Kingdom: A Study in Market Structure, Conduct, and Performance Analysis*. Nottingham: Department of Agricultural Economics, University of Nottingham.

Maxam, A. M., and W. Gilbert. 1977. A new method of sequencing DNA. *Proceedings of the National Academy of Sciences* 74: 560–4.

Merton, R. 1973. *The Sociology of Science*. Chicago: University of Chicago Press.

Meselson, M., and F. W. Stahl. 1958. The replication of DNA in *Escherichia coli*. *Proceedings of the National Academy of Sciences* 44: 671–82.

Miller, L. I. 1985. Biotechnology mergers signal industry consolidation. *Genetic Engineering News* 5 (2): 26.

Misawa, M. 1985. Production of useful plant metabolites. In *Advances in Biochemical Engineering/Biotechnology*, Plant Cell Culture, no. 31. Berlin: Springer Verlag, pp. 59–88.

Mitroff, I. 1974. *The Subjective Side of Science*. New York: Elsevier.

Mongelli, R. C. 1984. *Marketing Fresh Tomatoes: Systems and Costs*. Agricultural Marketing Service, Marketing Research Report, no. 1137. Washington, DC: USDA.

Monsanto. 1988. Farming: A picture of the future. *Science* 240: 1384.

Mooney, P. R. 1980. *Seeds of the Earth*. Ottawa, Inter Pares.

Morgan, D. 1979. *Merchants of Grain*. New York: Viking Press.

Morgan, T. H., A. H. Sturtevant, H. J. Muller, and C. B. Bridges. 1915. *The Mechanism of Mendelian Heredity*. New York: Holt, Rinehart, and Winston.

Morrison, G. 1938. *Tomato Varieties*. Special Bulletin 290. East Lansing: Michigan State College, Agricultural Experiment Station.

Moses, P. B., J. E. Tavares, and C. E. Hess. 1988. Funding agricultural biotechnology research. *Bio/Technology* 48: 215–17.

Moshy, R. 1986. Biotechnology: Its potential impact on traditional food processing. In S. K. Harlander and T. P. Labuza (eds), *Biotechnology in Food Processing*. Park Ridge, NJ: Noyes Publications, pp. 1–14.

Mulkay, M. 1977. Sociology of the scientific research community. In I. Spiegel-Rosing and D. de Solla Price (eds), *Science, Technology, and Society*. London: Sage Publications, pp. 93–148.

—— 1979. *Science and the Sociology of Knowledge*. London: George Allen and Unwin.

Muller, H. J. 1922. Variation due to change in the individual gene. *American Naturalist* 56: 32.

—— 1927. Artificial transmutation of the gene. *Science* 66: 84–7.

—— 1936. The need of physics in the attack on fundamental problems of genetics. *Scientific Monthly* 44: 210–14.

Murray, A. 1984. Japanese researchers report on enzyme engineering. *Bio/Technology* 2: 13–14.

Nair, K. 1979. *In Defense of the Irrational Peasant*. Chicago: University of Chicago Press.

Naj, A. K. 1989. Clouds gather over the biotech industry. *Wall Street Journal*, 30 Jan, B1, B5.

National Association of State Universities and Land-Grant Colleges, 1983. *Emerging Biotechnologies in Agriculture: Issues and Policies*. Progress Report II. Washington, DC: NASULGC, Division of Agriculture, Committee on Biotechnology.

—— 1984. *Emerging Biotechnologies in Agriculture: Issues and Policies*. Progress Report III. Washington, DC: NASULGC, Division of Agriculture, Committee on Biotechnology.

—— 1985. *Emerging Biotechnologies in Agriculture: Issues and Policies*. Progress Report IV. Washington, DC: NASULGC, Division of Agriculture, Committee on Biotechnology.

—— 1986. *Emerging Biotechnologies in Agriculture: Issues and Policies*. Progress Report V. Washington, DC: NASULGC, Division of Agriculture, Committee on Biotechnology.

—— 1989. *Emerging Biotechnologies in Agriculture: Issues and Policies*. Progress Report VIII. Washington. DC: NASULGC, Division of Agriculture, Committee on Biotechnology.

National Commission on Productivity. 1973. *Productivity in the Food Industry*. Washington, DC: US Government Printing Office.

National Institutes of Health. 1986. Recombinant DNA research: Actions under guidelines. *Federal Register* 51 (88): 16952–85.

National Research Council. 1972. *Report of the Committee on Research Advisory to the US Department of Agriculture*. PB 213 338. Washington, DC: National Technical Information Service.

—— 1984. *Genetic Engineering of Plants*. Washington, DC: National Academy of Sciences.

Nelkin, D. 1984. *Science as Intellectual Property*. New York: Macmillan.

Netzer, W. 1983a. Agrigenetics researchers express plant genes in tobacco plantlets. *Bio/Technology* 1: 461–2.

—— 1983b. IPRI applies for patent on cereal vector systems. *Genetic Engineering News* 3 (5): 11.

—— 1987a. Researchers seek ways to raise crop yields by improving photosynthesis. *Genetic Engineering News* 7 (10): 8, 70.

—— 1987b. Phosphate solubilizing genes might revolutionize fertilizer technology. *Genetic Engineering News* 7 (8): 10, 41.

Nevins, D. J. 1987. Why tomato biotechnology?: A potential to accelerate applications. In D. J. Nevins and R. A. Jones (eds), *Tomato Biotechnology*. New York: Alan R. Liss, pp. 3–14.

Nevins, D. J., and R. A. Jones. 1987. Preface. In D. J. Nevins and R. A. Jones (eds), *Tomato Biotechnology*. New York: Alan R. Liss, pp. xvii–xviii.

Newell, N., and S. Gordon. 1986. Profit opportunities in biotechnology for the food processing industry. In S. K. Harlander and T. P. Labuze (eds), *Biotechnology in Food Processing*, Park Ridge, NJ: Noyes Publications, pp. 297–311.

Newmark, P. 1989. UK biotech spending spree of a sort. *Bio/Technology* 7: 334.

Nirenberg, M. W., and J. H. Matthaei, 1961. The dependence of cell-free protein synthesis in *E. coli* upon naturally occurring or synthetic polyribonucleotides. *Proceedings of the National Academy of Sciences* 47: 1588–1602.

Norman, D. 1978. Farming systems research to improve the livelihood of small farmers. *American Journal of Agricultural Economics* 60: 813–18.

Norton, B. 1975. Biology and philosophy: The methodological foundations of biometry. *Journal of the History of Biology* 8: 85–93.

Oasa, E. K., and Swanson, L. E. 1986. The limits of farming systems research and development: Should development administrators be interested? *Agricultural Administration* 23: 201–21.

Office of Science and Technology Policy (OSTP). 1986. Coordinated framework for the regulation of biotechnology: Announcement of policy and notice for public comment. *Federal Register* 51 (123): 23302–50.

Office of Technology Assessment (OTA), US Congress. 1981. *An Assessment of the United States Food and Agricultural Research System.* Washington, DC: US Government Printing Office.

—— 1984. *Commercial Biotechnology: An International Analysis.* OTA-BA-218. Washington, DC: US Government Printing Office.

—— 1987. *New Developments in Biotechnology: Public Perceptions of Biotechnology.* OTA-BP-BA-45. Washington, DC: US Government Printing Office.

—— 1988. *New Developments in Biotechnology: U.S. Investments in Biotechnology.* OTA-BA-360. Washington, DC: US Government Printing Office.

—— 1989. *New Developments in Biotechnology: Patenting Life.* Washington, DC: US Government Printing Office.

Organisation for Economic Cooperation and Development (OECD). 1983. *The OCDE [sic] Scheme for the Application of International Standards for Fruits and Vegetables.* Paris: OECD.

—— 1988. *Biotechnology and the Changing Role of Government.* Paris: OECD.

Orr, D. W. 1988. Food alchemy and sustainable agriculture. *BioScience* 38 (December): 801–2.

Orr, T. 1985. DNA Plant Technology plots trajectory from lab to marketplace. *Genetic Engineering News* 5 (2): 14.

Orton, T. J. 1988. Biotechnology and the plant sciences. In W. Lacy and L. Busch (eds), *Biotechnology and Agricultural Cooperatives: Opportunities and Challenges.* Lexington, KY: Kentucky Agricultural Experiment Station, pp. 11–23.

Pauling, L. 1962. Early work on X-ray diffraction in the California Institute of Technology. In P. P. Ewald (ed.), *Fifty Years of X-Ray Diffraction.* Utrecht: Oosthock, pp. 623–8.

—— 1970. Fifty years of progress in structural chemistry and molecular biology. *Daedalus* 99 (4): 988–1014.

Pauling, L., and R. B. Corey. 1953. A proposed structure for the nucleic acids. *Proceedings of the the National Academy of Sciences* 39: 84–97.

Pauling, L., R. B. Corey, and H. R. Branson. 1951. The structure of proteins: Two hydrogen-bonded helical configurations of the polypeptide chain. *Proceedings of the National Academy of Sciences* 37: 205–11.

Pauly, P. J. 1987. *Controlling Life: Jacques Loeb and the Engineering Ideal in Biology.* New York: Oxford University Press.

Penn, J. B. 1979. The structure of American agriculture: Overview of the issues.

Structure Issues in American Agriculture. Agricultural Economic Report no. 438 Washington, DC: USDA, Economics, Cooperatives, and Statistics Service, pp. 2–23.

Perutz, M. F., M. G. Rossmann, A. F. Cullis, H. Muirhead, G. Will, and A. C. T. North. 1960. Structure of haemoglobin: A three-dimensional Fourier synthesis at 5.5-A. Resolution, obtained by X-ray analysis. *Nature* 185: 416–22.

Picard, E. 1988. Sélection du blé: l'intégration des biotechnologies. *Biofutur* 5: 48–58.

Pineiro, M., E. Trigo, and R. Fiorentino. 1979. Technical change in Latin American agriculture. *Food Policy* 4 (3): 169–77.

Plucknett, D., J. H. Nigel, J. H. Smith, J. T. Williams, and N. M. Anishetty. 1987. *Gene Banks and the World's Food.* Princeton: Princeton University Press.

Portas, C. A. M. 1987. Research and experimental development programmes for processing – a broad survey. *Acta Horticulturae* 200: 17–29.

Porte, W. S. 1952. *Commercial Production of Tomatoes.* Farmers' Bulletin, no. 2045. Washington, DC: USDA.

Preston, L. 1978. *Research in Corporate Social Performance and Policy I.* New York: JAI Press.

Pursel, V. G., C. A. Pinkert, K. F. Miller, D. J. Bolt, R. G. Campbell, R. R. Palmiter, R. L. Brinster, and R. E. Hammer. 1989. Genetic engineering of livestock. *Science* 244: 1281–7.

Quattrocchio, F., E. Benvenuto, R. Tavazza, L. Cuozzo, and G. Ancora. 1986. A study of the possible role of auxin in potato "hairy root" tissues. *Journal of Plant Physiology* 123 (2): 143–9.

Quine, W. V. O. 1953. *From A Logical Point of View.* Cambridge, MA: Harvard University Press.

Quisenberry, K. S., and L. P. Reitz. 1974. Turkey wheat: Cornerstone of an empire. *Agricultural History* 48 (1): 98–114.

Radin, D. N., H. M. Behl, P. Proksch, and E. Rodriguez. 1982. Rubber and other hydrocarbons produced in tissue culture of Guayule (*Parthenium argentatum*). *Plant Science Letters* 26: 301–10.

Randolph, S. R., and C. Sachs. 1981. The establishment of applied sciences: Agriculture and medicine compared. In L. Busch (ed.), *Science, Agriculture, and Development.* Totawa, NJ: Allanheld, Osmun, pp. 131–56.

Rasmussen, W. D. 1968. Advances in American agriculture: The mechanical tomato harvester as a case study. *Technology and Culture* 9 (10): 531–43.

Ratafia, M., and T. Purinton. 1988. World agricultural markets. *Bio/Technology* 6: 280–1.

Ratner, M. 1989. Crop biotech '89: Research efforts are market driven. *Bio/Technology* 7: 337–41.

Ravetz, J. 1971. *Scientific Knowledge and Its Social Problems.* New York: Oxford University Press.

Rick, C. M. 1978. The tomato. *Scientific American* 239 (2): 77–87.

—— 1980. Tomato linkage survey. *Tomato Genetic Cooperative Reports* 30: 2–17.

—— 1986. Germplasm resources in the wild tomato species. *Acta Horticulturae* 190: 39–47.

Rifkin, J. 1983. *Algeny.* New York: Viking Press.

Riley, R. 1975. Origins of wheat. In A. Spicer (ed.), *Bread: Social, Nutritional, and Agricultural Aspects of Wheaten Bread.* London: Applied Science Publishers, pp. 27–45.

Rockefeller Foundation, 1982. *Science for Agriculture.* Report of a workshop on Critical Issues in American Agricultural Research, Winrock International Conference Center, Morrilton, AR, 14–15 June. Washington, DC: The Rockefeller Foundation.

Rogers, E. M. 1983. *Diffusion of Innovations.* 3rd ed. New York: Free Press.

Rogers, E. M., and F. Shoemaker. 1971. *Communication of Innovations.* 2nd ed. New York: Free Press.

Rogoff, M., and S. L. Rawlins. 1987. Food security: A technological alternative. *BioScience* 37: 800–7.

Roll-Hansen, N. 1978. Drosophila genetics: A reductionist research program. *Journal of the History of Biology* 11: 159–210.

Ronk, R. 1986. Federal regulation of food biotechnology. In S. K. Harlander and T. P. Labuza (eds), *Biotechnology in Food Processing.* Park Ridge, NJ: Noyes Publications, pp. 29–35.

Rosenberg, C. 1967. Factors in the development of genetics in the United States: Some suggestions. *Journal of the History of Medicine and Applied Sciences* 22 (January): 27–46.

Rosenberg, N. 1982. *Inside the Black Box: Technology and Economics.* Cambridge: Cambridge University Press.

Rosevear, A., and C.A. Lambe. 1985. Immobilized plant cells. In *Advances in Biochemical Engineering/Biotechnology,* Plant Cell Culture, no. 31. Berlin: Springer Verlag, pp. 37–58.

Rosset, P. M., and J. H. Vandermeer. 1986. The confrontation between processors and farm workers in the midwest tomato industry and the role of agricultural research and extension. *Agriculture and Human Values* 3 (3): 26–32.

Rougeon, F., P. Kourilsky, and B. Mach. 1975. Insertion of rabbit β-globin gene sequence into *E. coli* plasmid. *Nucleic Acids Research* 2: 2365–78.

Russell, S. 1989. Genetically engineered tomato is juicier, resists spoilage. *Lexington Herald-Leader,* 6 September, D14.

Ruttan, V. W. 1980. Bureaucratic productivity: The case of agricultural research. *Public Choice* 35: 529–47.

—— 1982. *Agricultural Research Policy.* Minneapolis: University of Minnesota Press.

Sanders, J. H., and V. W. Ruttan. 1978. Biased choice of technical change in Brazilian agriculture. In H. P. Binswanger and V. W. Ruttan (eds), *Induced Innovation.* Baltimore: Johns Hopkins University Press, pp. 276–96.

Sanger, D. J., and A. R. Coulsen. 1975. A new method for determining sequences in DNA by primed synthesis with DNA–polymerase. *Journal of Molecular Biology* 94 (3): 444–8.

Schaeffer, G. W., M. D. Lazar, and P. S. Baenziger. 1984. Wheat. In W. R. Sharpe, D. A. Evans, P. V. Ammarito, and Y. Yamada (eds), *Handbook of*

Plant Cell Culture, vol. 2. New York: Macmillan, pp. 108–36.

Schaeffer, G. W., and F. T. Sharpe, Jr. 1983. Mutations and cell selections: Genetic variation for improved protein in rice. In. L. D. Owens (ed.), *Genetic Engineering: Applications to Agriculture*. Totawa, NJ: Rowman and Allanheld, pp. 237–54.

Schafer, W. (ed.) 1983. *Finalization in Science*. Dordrecht: D. Reidel.

Schaffner, W., E. Serfling, and M. Jasin. 1985. Enhancers and eukaryotic gene transcription. *Trends in Genetics* 1: 224–30.

Scheler, M. [1924] 1980. *Problems of a Sociology of Knowledge*. London: Routledge and Kegal Paul.

Schmeck, H. M., Jr. 1990. Skill and wisdom, as well as the "Bottom Line", must guide biotech's practitioners. *Genetic Engineering News* 10 (1): 4.

Schmidt, J. W. 1977. Wheat – its role in America's heritage. In M. Thorne (ed.), *Agronomists and Food: Contributions and Challenges*. Special Publication no. 30. Madison, WI: American Society of Agronomy, pp. 45–52.

Schmitz, A., and D. Seckler. 1970. Mechanized agriculture and social welfare: The case of the tomato harvester. *American Journal of Agricultural Economics* 52 : 569–77.

Schnake, L. D. 1982. *Calculation of white pan bread marketing spreads. Wheat Situation and Outlook*. WS-259. Washington: Economic Research Service, USDA.

Schrage, M., and N. Henderson, 1984. Biotech becomes a global priority. *Washington Post*, 17 December, A1, A10.

Schrodinger, E. 1945. *What is Life? The Physical Aspect of the Living Cell*. New York: Macmillan.

Schuh, G. E. 1986. Revitalizing land-grant universities. *Choices*. 2nd Quarter, 6–10.

Schwartz, B. and J. M. Love. 1987. European Community enlargement: Implications for US competitiveness in processed peaches and tomatoes. *Acta Horticulturae* 203: 313–20.

Scowcroft, W. R., S. A. Ryan, R. I. S. Brettel, and P. J. Larkin. 1984. Somoclonal variation: A "new" genetic resource. In J. H. W. Holden and J. T. Williams (eds), *Crop Genetic Resources: Conservation and Evaluation*. London: George Allen and Unwin, pp. 258–67.

Senior, P. 1986. Scale up of a fermentation process. In S. K. Harlander and T. P. Labuza (eds), *Biotechnology in Food Processing*. Park Ridge, NJ: Noyes Publications, pp. 249–57.

Shepard, J. F., D. Bidney, T. Bursby, and R. Kemble. 1983. Genetic transfer in plants through interspecific protoplast fusion. *Science* 219: 683–88.

Shetty, Y. 1979. New look at corporate goals. *California Management Review* 22 (2): 71–79.

Shine, J., P. H. Seeburg, J. A. Martial, J. D. Baxter, and H. M. Goodman. 1977. Construction and analysis of recombinant DNA of human chorionic somatomammotropin. *Nature* 270: 494–9.

Shull, G. H. 1909. A pure-line method in corn breeding. In *Annual Report of the American Breeders Association* 5: 51–9. Washington, DC: American Breeders Association.

Simon, R. 1986. Mail-order enzymes. *Forbes* 137 (11): 46.

Smith, H. O., and K. W. Wilcox. 1970. A restriction enzyme from *Hemophilus influenzae*. I. Purification and general properties. *Journal of Molecular Biology* 51: 379–91.

Smith, R. E. 1908. *Wheat Fields and Markets of the World*. St. Louis: Modern Miller.

Sneep, J., B. R. Murty, and H. F. Utz. 1979. Current breeding methods. In J. Sneep and A. J. I. Hendriksen (eds), *Plant Breeding Perspectives*. Wageningen, Netherlands: Centre for Agricultural Publishing and Documentation, pp. 104–233.

Spicer, A. 1975. The history of wheat and bread. In A. Spicer (ed.), *Bread: Social, Nutritional, and Agricultural Aspects of Wheaten Bread*. London: Applied Science Publishers, pp. 1–4.

Splinter, W. E. 1974. Harvest, handling, and storage. In G. E. Inglett (ed.), *Wheat: Production and Utilization*. Westport, CT: AVI, pp. 52–71.

Staba, J. 1985. Milestones in plant tissue culture systems for the production of secondary products. *Journal of Natural Products* 48: 203–9.

Stallman, J. 1986. Impacts of the 1930 Plant Patent Act on private fruit breeding investments. Ph.D. dissertation, Michigan State University.

Stallman, J., and Schmid, A. 1987. Property rights in plants: Implications for biotechnology research and extension. *American Journal of Agricultural Economics* 69: 432–37.

Stanley, D. W. 1986. Chemical and structural determinants of texture of fabricated foods. *Food Technology* 40 (3): 65–8, 76.

Steinberg, D. I., R. I. Jackson, K. S. Kim, and H. Song. 1981. *Korean Agricultural Research: The Integration of Research and Extension*. Washington, DC: Agency for International Development.

Stent, G. 1968. That was the molecular biology that was. *Science* 160: 390–95.

Sterling, J. 1988. Agbio products edge closer to market place. *Genetic Engineering News* 8 (5): 1, 15, 16.

Stern, C. 1970. The continuity of genetics. *Daedalus* 99 (4): 882–908.

Stevens, M. A. 1986. The future of the field crop. In J. G. Atherton and J. Rudich (eds), *The Tomato Crop*. London: Chapman and Hall, pp. 549–79.

Stevens, M. A., and C. M. Rick. 1986. Genetics and breeding. In J. D. Atherton and J. Rudich (eds), *The Tomato Crop*. London: Chapman and Hall, pp. 35–109.

Strange, M. 1988. *Family Farming: A New Economic Vision*. Lincoln, NB: University of Nebraska Press.

Stumpf, P. K. 1989. Agricultural research initiative. *Science* 244: 1029.

Sun, M. 1988. Designing food by engineering animals. *Science* 240: 240.

—— 1989. Market sours on milk hormone. *Science* 246: 876–7.

Sutton, S. B. 1977. How the department got its start. *Horticulture* 55 (4): 33–37.

Thompson, P. 1983–4. Risk, ethics and agriculture. *Journal of Environmental Systems* 13 (2): 135–153.

Thor, E., and J. W. Mamer. 1969. Rural manpower – overview. In B. F. Cargill and G. E. Rossmiller (eds), *Fruit and Vegetable Harvest Mechanization: Technological Implications*. East Lansing: Michigan State University, Rural Manpower Center, pp. 51–71.

Tigchelaar, E. C. 1986. Tomato breeding. In M. J. Bassett (ed.), *Breeding Vegetable Crops*. Westport, CT: AVI, pp. 135–71.

Tikoo, S. K. 1986. Breeding tomatoes for processing in India: Present status and future prospects. *Acta Horticulturae* 200: 73–81.

Tilghman, S. M., D. C. Tiermeier, J. G. Seidman, B. M. Peterlin, M. Sullivan, J. V. Maijel, and P. Leder. 1978. Intervening sequence of DNA identified in the structural portion of a mouse β-globin gene. *Proceedings of the National Academy of Sciences* 75: 725–9.

Timberlake, W. E., and M. A. Marshall. 1989. Genetic engineering of filamentous fungi. *Science* 244: 1313–25.

Timofeef-Ressovsky, N. W., E. G. Zimmer, and M. Delbruck. 1935. Uber die natur der genmutation und der genstrucktur. *Nachrrichten ans der Biologie der Gesellschaft der Wissenschaft zu Gottingen* 1: 189.

Timoshenko, V. P. [1932] 1972. *Agricultural Russia and the Wheat Problem*. Stanford, CA: Food Research Institute. Reprint. New York: Johnson Reprint Corporation.

Trigo, E. and M. E. Pineiro. 1982. Funding agricultural research. Paper presented at second meeting of National Agricultural Research Systems of Latin America and the Caribbean. Madrid.

—— 1983. Foundations of a science and technology policy for Latin American agriculture. In M. Pineiro and E. Trigo (eds), *Technical Change and Social Conflict in Agriculture: Latin American Perspectives*. Boulder, CO: Westview Press, pp. 165–73.

Tsai, D. H. and J. E. Kinsella. 1981. Initiation and growth of callus and cell suspensions of *Theobroma cacao L. Annals of Botany* 48: 459–557.

Tudge, C. 1988. *Food Crops for the Future*. Oxford: Basil Blackwell.

United States Bureau of the Census. 1985. 1982 census of wholesale trade. *Establishment and Firm Size*. WC82-I-1. Washington, DC: US Bureau of the Census.

—— 1986. 1982 census of manufacturing. *Concentration Ratios in Manufacturing*. MC82-S-7. Washington, DC: US Bureau of the Census.

United States Department of Agriculture (USDA). 1957. *Farm–Retail Price Spreads for Food Products*. Miscellaneous Publication no. 741. Washington, DC: USDA, Agricultural Marketing Service.

—— 1972. *Farm–Retail Price Spreads for Food Products*. Miscellaneous Publication no. 741 (revised). Washington, DC: USDA, Economic Research Service.

—— 1974. *Wheat Situation*. WS-230 Washington, DC: USDA, Economic Research Service.

—— 1976. *Wheat Situation*. WS-236. Washington, DC: USDA, Economic Research Service.

—— 1982a. *Developments in Farm to Retail Price Spreads for Food Products for 1981*. Agricultural Economic Report, no. 488. Washington, DC: USDA, Economic Research Service.

—— 1982b. *Economic Indicators of the Farm Sector: Income and Balance Sheet Statistics, 1981*. ECIFS. Washington, DC: USDA, Economic Research Service.

—— 1982c. *Manual of Classification of Agricultural and Forestry Research*.

Revision IV. Washington, DC: USDA, Current Research Information Service, Coooperative State Research Service.

—— 1983. *Agricultural Research Service Program Plan, 6 Year Implementation Plan, 1984–1990,* Washington, DC: USDA, Agricultural Research Service.

—— 1984a. *Wheat Outlook and Situation Report.* WS-269. Washington, DC: USDA, Economic Research Service.

—— 1984b. *Wheat: Background for 1985 Farm Legislation.* Agriculture Information Bulletin, no. 467. Washington, DC: USDA, Economic Research Service.

—— 1987. *Vegetable Situation and Outlook.* TVS-243. Washington, DC: USDA, Economic Research Service.

—— 1989. *Food Cost Review, 1988.* Agricultural Economic Report, no. 615. Washington, DC: USDA, Economic Research Service.

Uyeshiro, R. Y. 1977. *The Processing Tomato Industries of Greece, Portugal, and Spain.* Bulletin M-278. Washington, DC: USDA, Foreign Agricultural Service.

Van Brunt, J. and A. Klausner. 1987. The *Bio/Technology* roundtable on plant biotech. *Bio/Technology* 5: 128, 130–3.

van den Doel, K. and G. Junne. 1986. Product substitution through biotechnology: Impact on the Third World. *Trends in Biotechnology* 4: 88–90.

Vandermeer, J. H. 1982. Science and class conflict: The role of agricultural research in the midwest tomato industry. *Studies in Marxism* 12: 41–57.

—— 1986. Mechanized agriculture and social welfare: The tomato harvester in Ohio. *Agriculture and Human Values* 3 (3): 21–5.

Varrin, R. D., and D. S. Kukich, 1985. Guidelines for industry-sponsored research at universities. *Science* 227: 385–8.

Varro, M. T. 1912. *Varro on Farming.* Translated by L. Storr-Best. London: G. Bell and Sons.

Vasil, I. (ed.) 1987. *Biotechnology: Perspectives, Policies and Issues.* Gainesville, FL: University Presses of Florida.

Vergopoulos, K. 1985. The end of agribusiness or the emergence of biotechnology. *International Social Science Journal* 37 (3): 285–9

Villareal, R. L. 1980. *Tomatoes in the Tropics.* Boulder, CO: Westview Press.

Virgil. 1947. *The Georgics of Virgil.* Translated by C. Day Lewis. New York: Oxford University Press.

Vogeler, I. 1981. *The Myth of the Family Farm: Agribusiness Dominance of US Agriculture.* Boulder, CO: Westview Press.

Voorhees, E. B. 1898. *Tomato Growing.* Farmers' Bulletin, no. 76. Washington, DC: USDA.

Walsh, J. 1985. New R&D centers will test university ties. *Science* 227: 150–2.

Ward, B. 1989. Biology: The next frontier. *Sky* (2): 85.

Watson, J. D. 1968. *The Double Helix.* New York: Atheneum.

—— 1970. *Molecular Biology of the Gene.* Menlo Park, CA: Benjamin.

Watson, J. D., and F. H. C. Crick. 1953a. Molecular structure of nucleic acids: A structure for deoxyribonucleic acid. *Nature* 171: 737–8.

—— 1953b. Genetical implications of the structure of deoxyribonucleic acid. *Nature* 171: 964–7.

Watson, J. D. and J. Tooze. 1981. *The DNA Story: A Documentary of Gene Cloning*. San Francisco: W. H. Freeman.

Watson, J. D., J. Tooze, and D. T. Kuntz. 1983. *Recombinant DNA: A Short Course*. New York: W. H. Freeman.

Weiss, B., and C. C. Richardson 1967. Enzymatic breakage and joining of deoxyribonucleic acid. I. Repair of single-strand breaks in DNA by an enzyme system from *Escherichia coli* infected with T4 bacteriophage. *Proceedings of the National Academy of Sciences* 57: 1021–8.

Wheat, D. 1986. Strategies for commercialization of biotechnology in the food industry. In S. K. Harlander and T. P. Labuza (eds), *Biotechnology in Food Processing*. Park Ridge, NJ: Noyes Publications, pp. 279–84.

Wheeler, D. L. 1988. Harvard University receives first US patent issued on animals. *Chronicle of Higher Education* 34 (32): 1.

Wik, R. M. 1966. Science and American agriculture. In D. D. Vantassel and M. G. Hall (eds), *Science and Society in the United States*. Homewood, IL: Dorsey, pp. 81–106.

Wilkins, M. H. F., A. R. Stokes and H. R. Wilson. 1953. Molecular structure of deoxypentose nucleic acid. *Nature* 171: 738–40.

Winner, L. 1986. *The Whale and the Reactor*. Chicago: University of Chicago Press.

Witt, S. C. 1985. *Biotechnology and Genetic Diversity*. San Francisco: California Agricultural Lands Project.

—— 1990. *Briefbook: Biotechnology, Microbes and the Environment*. San Francisco: Center for Science Information.

Wittwer, S. H. 1986. New technology needed to sustain increased food production. In *Food for the Future*. Philadelphia: Philadelphia Society for the Promotion of Agriculture.

Worster, D. 1979. *Dust Bowl: The Southern Plains in the 1930's*. New York: Oxford University Press.

Wright, C. P. 1927. India as a producer and exporter of wheat. *Wheat Studies of the Food Research Institute* 3 (8): 317–41.

Wright, C. P., and J. S. Davis. 1925. Canada as a producer and exporter of wheat. *Wheat Studies of the Food Research Institute* 1 (8): 217–86.

Wysocki, B., Jr., 1987. Japanese now target another field the US leads: Biotechnology. *Wall Street Journal* 17 December, 1, 18.

Yamada, Y. 1984. Selection of cell lines for high yields of secondary metabolites. In I. K. Vasil (ed), *Cell Culture and Somatic Cell Genetics of Plants*, vol. 1. New York: Academic Press, pp. 629–36.

Yamada, Y. and Y. Fujita. 1983. Production of useful compounds in culture. In D. A. Evans, W. R. Sharp, P. V. Ammirato, and Y. Yamada (eds), *Handbook of Plant Cell Culture*, vol. 1. New York: Macmillan, pp. 717–28.

Yordanov, M. 1983. Heterosis in the tomato. In R. Frankel (ed.) *Monographs on Theoretical and Applied Genetics*. Vol. 6, *Heterosis*. Berlin: Springer-Verlag, pp. 189–219.

Yoxen, E. 1981. Life as a productive force: Capitalizing the science and technology of molecular biology. In L. Levidow and R. M. Young (eds), *Science, Technology, and the Labor Process*. London: CSE Books, pp. 66–122.

—— 1983. *The Gene Business: Who Should Control Biotechnology?* London: Pan Books.

Zenk, M. H. 1978. The impact of plant cell culture on industry. In T. A. Thorpe (ed.), *Frontiers of Plant Tissue Culture* Calgary: International Society for Plant Tissue Culture, pp. 1–13.

Zenk, M. H., M. Ruffer, T. M. Kutchan, and E. Gaineder. 1988. Biotechnological approaches to the production of isoquinoline alkaloids. In G. Bock and J. Marsh (eds), *Applications of Plant Cell and Tissue Culture*. Ciba Foundation Symposium, no. 137. Chichester: John Wiley and Sons, pp. 213–27.

Zhou, G., J. Weng, J. Huang, S. Qian, and Q. Liu. 1983. Exogenous DNA Caused Phenotypic Variation in Cotton. *12th Annual UCLA Symposium on Plant Molecular Biology. Journal of Cell Biology, Supplement* 7 (Part B): 250.

Zwart, A. C., and K. D. Meilke. 1979. The influence of domestic pricing policies and buffer stocks on price stability in the world wheat industry. *American Journal of Agricultural Economics* 61: 434–47.

Index

DATE DUE

APR 1 4 199			

HIGHSMITH 45-102   PRINTED IN U.S.A.